AVIAN
ARCHITECTURE

AVIAN ARCHITECTURE

HOW BIRDS DESIGN, ENGINEER & BUILD
PETER GOODFELLOW

Consultant editor Mike Hansell

Ivy Press

First published in Great Britain in 2013 by

Ivy Press
210 High Street
Lewes, East Sussex
BN7 2NS, UK
www.ivypress.co.uk

British Library Cataloguing-in-Publication Data
A CIP catalogue record for this book is available
from the British Library.

ISBN 978-1-908005-84-7

Ivy Press
This book was conceived, designed
and produced by Ivy Press
CREATIVE DIRECTOR Peter Bridgewater
PUBLISHER Jason Hook
EDITORIAL DIRECTOR Caroline Earle
ART DIRECTOR Michael Whitehead
COMMISSIONING EDITOR Kate Shanahan
ASSISTANT EDITOR Jamie Pumfrey
PICTURE MANAGER Katie Greenwood
DESIGN Martin Topping
ILLUSTRATOR Coral Mula

Typeset in Scala

Colour origination by Ivy Press Reprographics
Printed in China

10 9 8 7 6 5 4 3 2 1

Distributed worldwide (except North America)
by Thames & Hudson Ltd., 181A High Holborn,
London WC1V 7QX, United Kingdom

The cedars of Lebanon get plenty of rain –
the Lord's own trees, which he planted.
There the birds build their nests;
the storks nest in the fir trees...
THE GOOD NEWS BIBLE, PSALM 104

**For my grandchildren, who help to keep me young –
Katherine and Scott, and Emmie and Kai.**

Front cover photo: Photolibrary/Donal Mullins

Contents

Foreword
by Professor Mike Hansell

We humans are great builders, so we tend to admire other animals that build. The most consistently excellent builders among the vertebrate animals are the birds. That it should be the birds is somewhat surprising. Why not our nearest living relatives, the mammals? The truth is that we can think of any number of examples of well-built, even intricate bird nests, but few comparable structures built by mammals.

The excellence of bird-nest building is surprising for a second reason. Birds generally spend very little time in nest building. Overwhelmingly, nests serve as containers for eggs. Many of them do also provide a secure home for the growing chicks, but only in rare cases are nests used as dormitories outside the breeding season. So, in most cases, the working life of a nest is only a few weeks. Nests may be elegantly constructed, but to do a job that is generally brief.

The third surprise about bird nests is how the birds build them. A bird's building equipment is largely just the beak. To build even a simple nest with just a beak would seem to be a bit like trying to make a ham and cheese sandwich with one hand behind your back. However, birds have the advantage of very flexible necks and good vision; what is more, the beak does not work entirely on its own. The feet of some species are important for scratching, and occasionally for holding nest material. More unexpectedly, the rounded belly or breast of a bird is often used to mould the interior shape of the nest. The results achieved by birds using this combination

ABOVE
CLIFF SWALLOW MUD NEST
An American Cliff Swallow (Petro-chelidon pyrrhonota) guards its cave-like construction of mud and fibres.

RIGHT
HUMMINGBIRD CUP NEST
The carefully crafted insulation of plant down is clearly visible in this cosy and durable cup nest.

of equipment are nests that are frequently complex and often beautiful. This raises another question about the building process: how do birds know what to do?

Most of the evidence we have points to a strong genetic influence on the way birds build. When a bird becomes adult, it can build a species-typical nest without apparent practice. The presumption, therefore, has tended to be that birds do not need to learn nest building at all. If true, this is very surprising, because in all their other activities that so impress us (migration, food finding and song) birds show considerable learning skills that modify and extend their innate abilities. It may be that, were we to look closer at bird-nest building, we would be able to see that experienced birds do make better nests than first-timers. Certainly, in the case of weaverbirds, we know that this is the case, although how and what they learn remains to be studied.

The strongest evidence for the role of learning comes not from nest building but from the construction of the elaborate and colourfully ornamented display structures of male bowerbirds. Males in some of these species apparently need several years of study and practice to become skilful enough bower builders to attract females.

The nests and other structures that birds build are varied in size, materials and design. They deserve to be better known and celebrated. This richly illustrated book is an opportunity to explore this wonderful variety and complexity.

Introduction

The book is divided into thematic chapters, each of which explores the architectural characteristics and individual variations of a particular type of nest. The evolutionary pattern of nest building is debated, but it is clear that nest evolution can be very rapid – as demonstrated by the different forms that can be found within one family. The nest is dependent upon, and adapted to, the habitat in which birds attempt to survive and reproduce. While this work cannot claim to be encyclopaedic, it can explore how different species have evolved common architectural and engineering techniques to adapt their nests to the substrate and available materials. It is the architectural ability of birds to build a variety of nest types that has enabled them to diversify into so many habitats – from the desert to the Antarctic, from high in trees to underground, from open ground to out on the water – and which creates some of the best engineered structures in the natural world.

How the book works

Each chapter begins with an overview of a specific nest type, outlining key structural characteristics and building methods, and identifying the varied bird families and species that construct a specific type of nest. The different families associated with a nest type will often share little or nothing else in common. They may vary greatly in size, habitat, courtship and rearing behaviour; and yet, in the matter of nest construction they find common ground.

Blueprints

Following each chapter's introduction, the architectural characteristics of specific forms of the nest type are presented as 'Blueprint' drawings. These annotated illustrations show the structure, shape and dimensions of archetypal nests, while also highlighting specific architectural elements. In addition to providing a unique perspective on nest construction, and suggesting variations within the nest type, the Blueprints also help to place the nest type in the context of different habitats.

Materials and features

The Materials and Features pages each offer a close-up study of the nest of an individual species. Illustrations and accompanying photographs depict unique features. These include camouflage; how construction has been adapted to habitat; characteristic and occasional materials; and, where appropriate, how we have imitated nature in our own architecture – a process known as biomimicry.

Building techniques

The Building Techniques pages study unique and remarkable construction skills – such as the stitching and weaving of some passerines – through step-by-step illustrations. They are, as such, a celebration of the diverse ingenuity and great dexterity of the finest of avian architects and builders.

Case studies

The Case Study pages of each chapter provide examples of how different species adapt the nest type to their specific habitat and requirements. The key characteristics of their nest type are described alongside general information on the species, and details of nest location, building techniques and materials. Nests are primarily built to rear young rather than as permanent homes, so there are also notes on courtship and mating, monitoring of eggs and care for the young. Some case studies also feature step-by-step illustrations highlighting a bird's distinctive architectural behaviour and building techniques.

The book also features those birds that have a nest site but do not construct a nest; for instance, the White Tern (*Gygis alba*) lays a single egg in a crack on a horizontal branch or rocky ledge. Finally, there are those species that build structures other than nests. A book on avian architecture would be incomplete without the ornate constructions of the bowerbirds of Australia and New Guinea.

ABOVE
SOUTHERN MASKED WEAVER
The Southern Masked Weaver
(Ploceus velatus) *exhibits one of the*
most intricate nest-building techniques.

CHAPTER ONE

Scrape Nests

The builders of scrape nests are the minimalists of the avian architecture fraternity. Faced with the challenge of an open habitat with limited materials, they literally scratch out an existence by gouging a shallow nest out of the ground. The scrape nest is exactly as it sounds: a scrape or depression in the earth, sometimes with material added to create a lining. Primarily an area for eggs and their incubation, the ground-level scrape offers quite limited defences. Camouflage is therefore essential, and the scrape nest, eggs and young are adapted to blend with the ground.

To start constructing a scrape, usually the hen will lower herself onto her breast at the chosen site and rotate and shuffle with her feet to form a shallow depression in the sand, shingle or vegetation. Many birds line their scrapes, the amount of lining varying with the species and individual, and two building methods are used. First, as a pair moves away from or around the scrape site they pick up nest material – bits of vegetation, small pebbles, shell fragments – and use a technique called 'sideways throwing' to toss the objects to the side of or beneath them. Second, the sitting bird pulls at the material and tucks it alongside or beneath her – this is 'sideways building'.

Some of the simplest scrapes are those of shore-nesting plovers. Wildfowl make more elaborate scrapes and add a layer of down. The Common Eider (*Somateria mollissima*) builds in the shelter of rock or vegetation and plucks feathers and down from her breast to establish a thick lining.

Despite its simplicity, the scrape nest demonstrates a degree of precision to effectively shelter the eggs and incubating bird, especially in cold, damp habitats. The scientist Jane Reid and her colleagues found that the scrape of the Arctic-breeding Pectoral Sandpiper (*Calidris melanotos*) was made to an optimal cup depth to result in the minimum heat loss for the eggs; too deep and the cold ground affected incubation, too shallow and wind chill was a problem.

Vulnerable to predators, the eggs in scrape nests demonstrate excellent camouflage. The eggs of sandy-shore species such as plovers generally have a light base colour, finely speckled with grey or black. Waders that nest on grasslands, tundra or marshes lay eggs with a darker base colour, as befits the nest site. Ducks' eggs are plainly coloured and rely on being hidden by undergrowth and down.

Various features aid the chicks' survival. They often leave the nest soon after hatching, and are precocial (able to walk almost immediately), and nidifugous ('nest flyers' – they flee the nest to seek food, guided by the parents). Their camouflaged plumage protects them when a parent's alarm call makes them crouch stock still.

RIGHT
PIPING PLOVER SCRAPE NEST
The shallow scrape and camouflaged eggs can be seen in the shingle-beach scrape of this Piping Plover (Charadrius melodus).

BLUEPRINTS

Scrape Nest Structures

The architectural blueprint for a scrape nest includes the shallow indentation on a ground site; precise depth to achieve optimal egg temperature; simple lining materials; and strong camouflage features. Species with scrape nests include game birds (pheasants, grouse, partridges); ducks; waders (shore-nesting plovers); the Short-eared Owl (*Asio flammeus*); and the Ostrich (*Struthio camelus*).

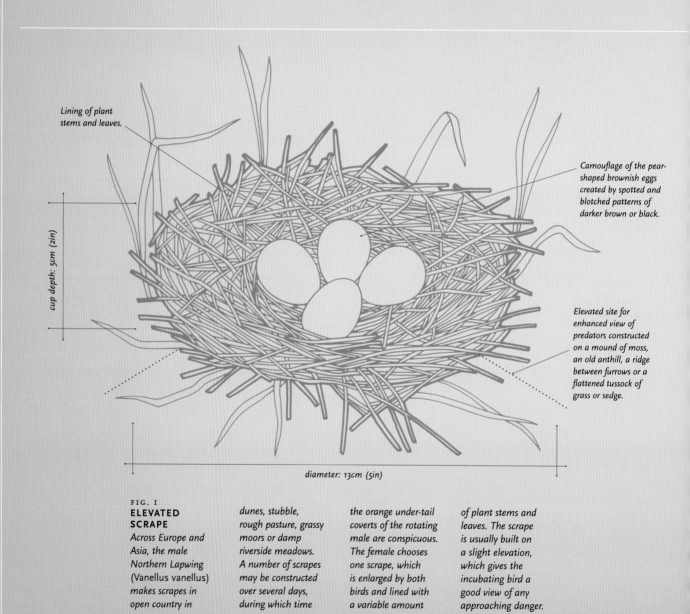

Lining of plant stems and leaves.

Camouflage of the pear-shaped brownish eggs created by spotted and blotched patterns of darker brown or black.

Elevated site for enhanced view of predators constructed on a mound of moss, an old anthill, a ridge between furrows or a flattened tussock of grass or sedge.

cup depth: 5cm (2in)

diameter: 13cm (5in)

FIG. I
ELEVATED SCRAPE
Across Europe and Asia, the male Northern Lapwing (Vanellus vanellus) makes scrapes in open country in dunes, stubble, rough pasture, grassy moors or damp riverside meadows. A number of scrapes may be constructed over several days, during which time the orange under-tail coverts of the rotating male are conspicuous. The female chooses one scrape, which is enlarged by both birds and lined with a variable amount of plant stems and leaves. The scrape is usually built on a slight elevation, which gives the incubating bird a good view of any approaching danger.

FIG. I NORTHERN LAPWING NEST

VARIETIES OF STRUCTURE

Sites include open ground, boggy ground hidden by growing vegetation, and slightly elevated platforms. The Short-eared Owl makes a scrape in heather moors, tall grass, dead reeds and marram grass on dunes. Scrapes are sometimes lined with materials including plant stems, leaves, grass, shell fragments and pebbles. Eggs may be camouflaged or buried for protection.

FIG. 2
HIDDEN SCRAPE
The Arctic-breeding Red-necked Phalarope (Phalaropus lobatus) constructs a scrape on boggy or marshy ground by pools or water-filled ditches. The nest becomes more hidden as the vegetation grows. Both sexes make several scrapes together, one of which is chosen by the female. The cup-shaped depression is lined with leaves, dry grasses and growing stems pulled over into the depression. After laying the eggs, the female has nothing more to do with the family. The male incubates the four well-camouflaged eggs and cares for the chicks.

FIG. 3
OPEN SCRAPE
The Kentish Plover (Charadrius alexandrinus) breeds on five continents but is declining wherever there is human disturbance. It nests on open shingle, sand or dry mud by the sea, lagoons or riversides. The male makes several scrapes from which the female chooses one. The shallow scrape may then be lined with shell fragments, bits of vegetation and small pebbles. In sand, the eggs may be half or almost wholly buried, with points downward.

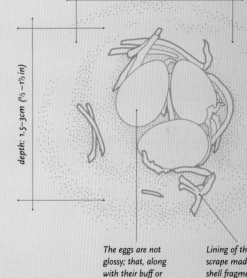

diameter: 6.5–10cm (2½–4in)

depth: 1.5–3cm (²⁄₃–1¹⁄₃ in)

The eggs are not glossy; that, along with their buff or sandy base colour with a scattering of black spots or streaks, helps them to hide on the sand or shingle.

Lining of the open scrape made with shell fragments and vegetation.

Concealment offered by growing grass drawn down over the scrape.

Lining of leaves and dry grasses.

depth: 2–4cm (³⁄₄–1¹⁄₃ in)

Camouflaged eggs and an incubating male that is not easily disturbed enhance the scrape's security.

diameter: 6.5–10cm (2½–4in)

FIG. 2 RED-NECKED PHALAROPE NEST | FIG. 3 KENTISH PLOVER NEST

MATERIALS AND FEATURES
Courser Nest

The coursers form a special subfamily of waders that has evolved into a group of eight species. They inhabit semi-desert or other almost bare ground in Africa and India. The most widespread is the Cream-coloured Courser (*Cursorius cursor*), which is found in barren country bordering the north and south of the Sahara in North Africa. All except one species are mostly sandy coloured. The courser's scrape nest is little more than a scratching in the sand, and normally two eggs are laid directly onto the ground. This minimalist architecture in fact provides the nest's primary defence, because the lack of structure together with the superb camouflage of eggs and chicks can make it virtually invisible. If it is disturbed, the courser has long, strong legs that enable it to run well (*cursor* means 'runner' in Latin). It runs in a hunched manner, then stops suddenly and stands tall, with neck stretched up for a view of the intruder.

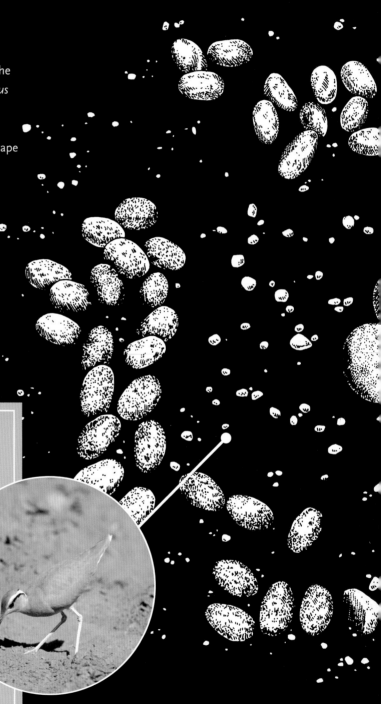

Invisibility

The courser nest uses camouflage as its primary defence. The nest structure is inconspicuous, and the eggs are camouflaged by a pale buff colour finely spotted with brown. When the adult lands, as British Museum scientist David Bannerman wrote, 'the bird will almost fade from view', so closely does it match the ground colour. The chicks are even better camouflaged: their pale sandy-rufous down is lightly speckled with white and grey, and they lack the adults' striking head pattern. The combined effect is a cloak of invisibility.

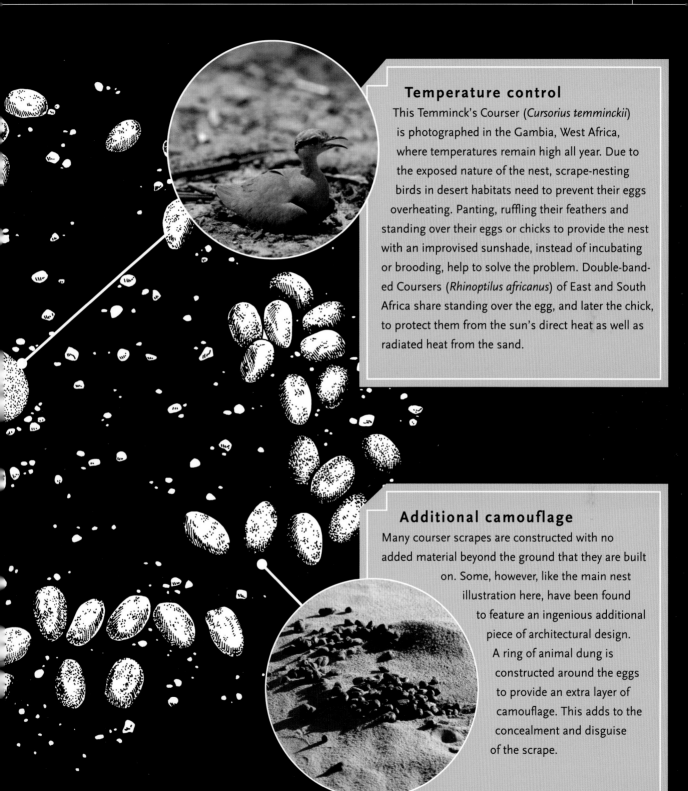

Temperature control

This Temminck's Courser (*Cursorius temminckii*) is photographed in the Gambia, West Africa, where temperatures remain high all year. Due to the exposed nature of the nest, scrape-nesting birds in desert habitats need to prevent their eggs overheating. Panting, ruffling their feathers and standing over their eggs or chicks to provide the nest with an improvised sunshade, instead of incubating or brooding, help to solve the problem. Double-banded Coursers (*Rhinoptilus africanus*) of East and South Africa share standing over the egg, and later the chick, to protect them from the sun's direct heat as well as radiated heat from the sand.

Additional camouflage

Many courser scrapes are constructed with no added material beyond the ground that they are built on. Some, however, like the main nest illustration here, have been found to feature an ingenious additional piece of architectural design. A ring of animal dung is constructed around the eggs to provide an extra layer of camouflage. This adds to the concealment and disguise of the scrape.

CASE STUDY
Killdeer

The scrape of the Killdeer uses camouflage to disguise a shallow nest on open ground. Found widely across North America, except Alaska, from southern Canada southward, the Killdeer winters in the southern states and as far south as northern South America. The Killdeer uses noise and display to distract intruders when the nest is threatened.

Classification

ORDER	Charadriiformes
FAMILY	Charadriidae
SPECIES	*Charadrius vociferus*
RELATED SPECIES	Other plovers, sandpipers, curlews
NEST TYPE	Scrape
SPECIES WITH SIMILAR NESTS	Terns, game birds, nightjars
NEST SPECIALIZATION	Minimal nest material

Habitat and nest

The Killdeer nests in a variety of sites where grass is short or absent – farm fields, golf courses, city parks, sports fields, roadsides, airports and even gardens. The nest is usually sited on a bare, sandy or gravelly area, and the Killdeer's four eggs are laid pointed ends inward in a shallow scrape or depression. The scrape is commonly unlined, or sparsely lined with a few plant fragments, wood chips or pebbles, all of which are gathered nearby.

Nest defence

The Killdeer offers an excellent example of how some ground-nesting species have adapted to the problem of a nest with visible eggs or chicks. Both parents are particularly noisy when disturbed by people or animals. If the incubating bird sees an advancing intruder, it will slip off the eggs, run and then give the alarm. If surprised, the bird will flounder and stagger away, beating the ground with its wings, calling madly in order to attract the attention of the potential predator. This technique is the 'broken-wing' trick. When the intruder has been led far enough away, the bird suddenly flies free to join its equally noisy partner. In its haste to flee, the bird occasionally disturbs the eggs; on its return the incubating bird rearranges them before it settles down.

Eggs and young

The male and female share the 24-day incubation of the eggs. The chicks are precocial, leaving the scrape soon after hatching. Their parents guard their young and lead them to insect-rich feeding grounds so they can feed themselves.

**KILLDEER
SHINGLE SCRAPE**
The Killdeer's simple depression in the scrape hardly contains the four eggs. However, their shape and mottled colouring, together with the pattern of the bird's plumage, camouflage them well on this gravelly site.

1. The Killdeer staggers with both wings flailing.

2. The left or right wing is dragged along the ground to mimic injury.

3. With the predator a safe distance from the scrape, the Killdeer flies back to the eggs.

ABOVE
DISTRACTION DISPLAY
On detecting a potential predator, the Killdeer will perform an elaborate 'broken-wing' distraction display, involving staggering forward, with both wings flailing (1.); then continuing to move away from the eggs, dragging either the left or right wing on the ground (2.) and (3.). This attracts the attention of the predator and lures it a safe distance from the eggs. With the threat removed, the Killdeer flies back.

CASE STUDY
Ostrich

The nest of the Ostrich answers the challenge of protecting eggs in a harsh environment not architecturally but socially. The male commonly forms a harem with three hens to maximize the number of eggs, all laid in one simple pit-like scrape. This improves the chance of a successful hatching. The Ostrich is found south of the Sahara, north of the tropical forest belt and in southwest Africa.

Habitat and nest

Ostriches live in a wide variety of habitats – semidesert, open savanna and dry wadis, avoiding ground where trees and shrubs are frequent. The nest is, therefore, created on open ground. It consists of a shallow depression scraped mostly by the male. Its overall diameter is about 2.7m (9ft), and it is 30–61cm (1–2ft) deep.

The shared nest

Ostriches are social birds throughout the year, sometimes in flocks of up to several dozen. Within each flock are dominant males and females. In the breeding season, a dominant female (the major hen) will initiate pair formation with a dominant male. She will allow minor females to join the harem, and all of the hens lay in the one nest. The major hen usually lays 4–8 eggs first; then the other females add to the clutch until there are about 20–25 eggs in total. She drives the minor hens from the scrape once they have laid.

Nest defence

It's normal for the major hen to incubate by day and the male by night. The nest is usually so clearly in the open that the daytime incubating bird often lies with its neck stretched out along the ground in order to be less conspicuous. The chicks begin calling one or more days before hatching, thus establishing close contact with their parents, who care for them for up to a year after hatching. To protect the chicks from predators, the male in particular has an elaborate distraction display, flapping its wings, dropping to the ground and running about erratically.

Classification

ORDER	Struthioniformes
FAMILY	Struthionidae
SPECIES	*Struthio camelus*
RELATED SPECIES	One close relative, the Somali Ostrich *Struthio molybdophanes*; other ratites are the Emu, cassowary and rhea
NEST TYPE	Scrape
SPECIES WITH SIMILAR NESTS	Emu, rhea
NEST SPECIALIZATION	Nest used by several females

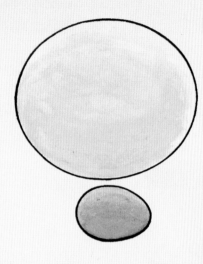

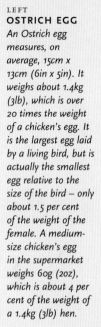

OSTRICH EGG
An Ostrich egg measures, on average, 15cm x 13cm (6in x 5in). It weighs about 1.4kg (3lb), which is over 20 times the weight of a chicken's egg. It is the largest egg laid by a living bird, but is actually the smallest egg relative to the size of the bird – only about 1.5 per cent of the weight of the female. A medium-size chicken's egg in the supermarket weighs 60g (2oz), which is about 4 per cent of the weight of a 1.4kg (3lb) hen.

1. Ostrich and egg.

2. Ostrich and chicken egg.

3. Chicken and egg.

OSTRICH SCRAPE
Despite the large size of the ostrich scrape, it is not uncommon for a large clutch from the harem to be too big for the incubating major hen to cover and incubate. Over 50 eggs in one nest have been recorded. Eggs of minor females are often discarded by the dominant female; this may help to ensure that her eggs have a better chance of survival. Many eggs fail to hatch. On average, only one chick survives to adulthood, which illustrates the vulnerability of ground nesting.

CASE STUDY
Arctic Tern

Constructed in tightly packed groups, the scrape nests of the Arctic Tern demonstrate the defensive capability afforded by nest building in colonies. A summer visitor to the Northern Hemisphere, the Arctic Tern breeds mostly within the Arctic Circle, but also as far south as the British Isles, and Labrador, Nova Scotia and Massachusetts in North America. They winter at sea in the Southern Ocean.

Classification

ORDER	Charadriiformes
FAMILY	Sternidae
SPECIES	*Sterna paradisaea*
RELATED SPECIES	Gulls and skuas
NEST TYPE	Scrape
SPECIES WITH SIMILAR NESTS	Other 'sea terns'
NEST SPECIALIZATION	Colony defence

Habitat and nest

The Arctic Tern's scrape is constructed on a grass-covered islet, a sandy or shingle beach, or even inland on heathland or tundra. The nest is shallow, with or without a little lining. Both sexes work, often alternately, to scrape the depression, in which two eggs are commonly laid.

Breeding

Arctic terns do not breed until they are at least three years old. They breed in solitary pairs, small groups and often dense, large colonies, the largest of which may number hundreds or even thousands of pairs. In the case of the largest colonies, the scrape nests are likely to be only a few metres apart. The territory of each pair then comprises only the roughly circular area around the nest. This space is just large enough for the pair's courtship and greeting displays, the scrape and the young.

Nest defence

Arctic Terns are aggressive in defence of their vulnerable ground nests, particularly when they have chicks, and here the colony of nests affords mutual protection. On sight of a predator, the colony will fly up in a 'dread', often started by birds in the densest part. Neighbours and wandering chicks are driven away. Predators that may have discovered the nests' location are mobbed and chased. These include gulls, skuas and crows. Peregrines (*Falco peregrinus*) and White-tailed Eagles (*Haliaeetus albicilla*) are mobbed aggressively. Animals and humans are 'dive-bombed' with excreta, and even struck with the bill, in defence of the colony's nests.

CASE STUDY
Common Eider

The scrape nest of the Common Eider is lined with a bed of the female's down. Eiders nest together for protection and the ducklings are cared for in groups. Breeding along the northern coasts of North America, the shores of Greenland and the coasts of northwest Europe, the Common Eider generally nests by the shore or on offshore islands.

Classification

ORDER	Anseriformes
FAMILY	Anatidae
SPECIES	*Somateria mollissima*
RELATED SPECIES	King Eider, Steller's Eider, Spectacled Eider
NEST TYPE	Hollow in the ground, well lined
SPECIES WITH SIMILAR NESTS	Most other wildfowl
NEST SPECIALIZATION	Thick lining of down

Habitat and nest

In the breeding season, the Common Eider is usually colonial, with as many as a few thousand nesting together, often as closely as two nests per square metre. The female builds the scrape, often in the shelter of a rock or well hidden in thick herbage. On coasts, the nests can be sited on turf or in a dune, where they are more exposed.

Nest lining

The nest is thickly lined with grass, seaweed and an abundant amount of the female's own breast feathers and down. The down feathers insulate the wintering bird at sea, but the female is able to sacrifice some of her own insulation because it is spring and the weather is improving. The eggs are also covered with a layer of down when the duck leaves the nest. Old Common Eider nests are reused, causing their architecture to develop into a permanent cup.

Eggs and young

Drakes guard their incubating mates but take no further part in family life. The ducklings hatch after four weeks and are cared for by the female, who leads them to a feeding area, usually near the nest. The ducklings leave the comfort of the nest as soon as they are dry, and are able to feed themselves. At this point the parent-young relationship weakens as the adults go off to feed and gain strength after losing weight during incubation. The ducklings, meanwhile, join those of other families, often forming groups of a hundred or more, which are tended by several ducks known as 'aunties'. The ducklings become independent after eight or nine weeks.

CHAPTER TWO

Holes & Tunnels

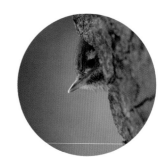

Hole and tunnel nesting answers the challenge of nest defence by exploiting the relative safety of such refuges as tree holes and rock crevices. Such nests are widespread, constructed by species from within some 50 per cent of bird orders. In addition to providing protection from predators, holes and tunnels also offer shelter from the weather, allowing the nesting bird to conserve energy. Not surprisingly, such prime architecture is the subject of intense competition, and it is not uncommon for latecomers to aggressively hijack the new build of another bird or species.

There are two forms of avian architect in this category. Primary cavity nesters construct their own nests, which can be a time-consuming and laborious undertaking. Secondary cavity nesters depend on natural or existing holes and tunnels. These are often inherited, or stolen, from primary cavity nesters, but older nests may be of poor quality or infested with parasites from previous use.

The true artists and engineers of hole and tunnel nesting are the primary cavity nesters, such as the kingfishers and woodpeckers. They exhibit beaks or bills that have evolved to form splendid building tools. The flattened bills of the woodpeckers are perfect for chiselling into tree bark; bee-eaters and jacamars have adapted slender, pointed beaks to dig effectively; and trogons employ short, stumpy bills to gouge their architectural surfaces. Those parrots that do make their own nest holes have similarly stunted beaks, which they use with a biting action during their excavations.

For a primary cavity nester, constructing a nest site is a major undertaking, and tropical species can begin a nest weeks before it is needed. They excavate in the rainy season when the soil or sand is soft, and rear their young when the rains have stopped.

Secondary cavity nesters are not mere squatters, and demonstrate architectural dexterity to varying degrees. They include the Great Tit (*Parus major*) of Eurasia, which builds a nest thickly lined with plant and other material in existing tree holes or man-made nest boxes. The Great Hornbill (*Buceros bicornis*) of Asia finds a suitable cavity and customizes it to its own specifications. An unexpected hole nester is the diving duck, the Common Goldeneye (*Bucephala clangula*). Unusually for an aquatic species, it will nest in a natural hole or crevice, or take over the hole made by a large woodpecker. Likewise, the Burrowing Owl (*Athene cunicularia*) takes over an existing ground burrow instead of making its own.

Holes and tunnels incorporate such security features as enclosed entrances and long, sloping tunnels, but snakes and small, agile mammals, such as weasels, ensure that the principal cause of nest failure is predation.

RIGHT
**BLACK-CAPPED
CHICKADEE HOLE**
*The Black-capped Chickadee (Poecile
atricapillus) pair will excavate a
natural cavity if the wood is rotten,
or adopt a woodpecker's old nest.*

BLUEPRINTS

Hole & Tunnel Structures

For hole and tunnel nests, the architectural blueprint takes the form of entrance, tunnel and egg chamber. The key division is between an excavated site (primary cavity nesters such as kingfishers, jacamars and woodpeckers) or found site (secondary cavity nesters such as ducks, trogons and hornbills). Only one wader is a hole nester, the Crab-plover (*Dromas ardeola*) of the Red Sea and Indian Ocean. Puffbirds, in the same order as toucans and woodpeckers, nest in holes. In Central America, the White-whiskered Puffbird (*Malacoptila panamensis*) forms its nest in a burrow.

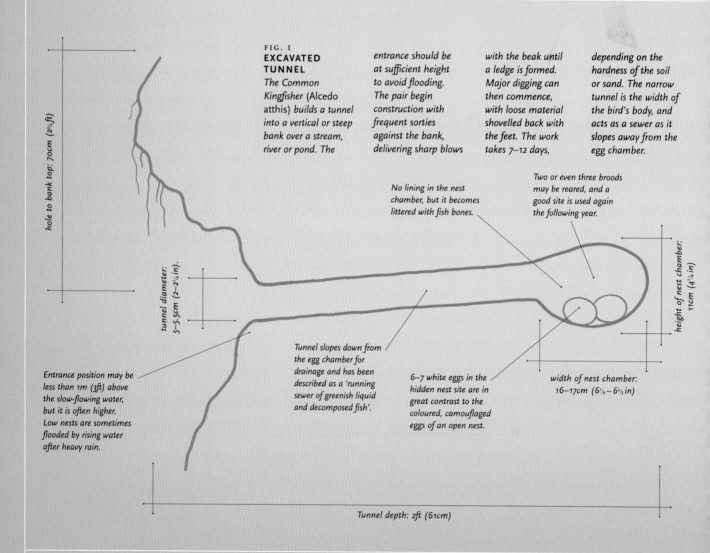

FIG. I
EXCAVATED TUNNEL
The Common Kingfisher (Alcedo atthis) builds a tunnel into a vertical or steep bank over a stream, river or pond. The entrance should be at sufficient height to avoid flooding. The pair begin construction with frequent sorties against the bank, delivering sharp blows with the beak until a ledge is formed. Major digging can then commence, with loose material shovelled back with the feet. The work takes 7–12 days, depending on the hardness of the soil or sand. The narrow tunnel is the width of the bird's body, and acts as a sewer as it slopes away from the egg chamber.

hole to bank top: 70cm (2¼ ft)

tunnel diameter: 5–5.5cm (2–2¼ in).

No lining in the nest chamber, but it becomes littered with fish bones.

Two or even three broods may be reared, and a good site is used again the following year.

height of nest chamber: 11cm (4¼ in)

Entrance position may be less than 1m (3ft) above the slow-flowing water, but it is often higher. Low nests are sometimes flooded by rising water after heavy rain.

Tunnel slopes down from the egg chamber for drainage and has been described as a 'running sewer of greenish liquid and decomposed fish'.

6–7 white eggs in the hidden nest site are in great contrast to the coloured, camouflaged eggs of an open nest.

width of nest chamber: 16–17cm (6¼–6⅔ in)

Tunnel depth: 2ft (61cm)

FIG. I COMMON KINGFISHER NEST

VARIETIES OF STRUCTURE

Primary cavity nesters excavate holes in trees and some other plants, and construct tunnels in the ground, sometimes in riverbanks. A number of species also construct cavities in termite mounds. Secondary cavity nesters adopt existing nest sites in living trees or rotten stumps. Some birds take over the deserted ground burrow of a mammal and convert it into a nest. Temperature is an essential consideration. In cold regions the nest may be lined for warmth. In hot desert, nest entrances are aligned to protect the eggs from heat.

FIG. 3
EXCAVATED CACTUS
The Gila Woodpecker (Melanerpes uropygialis), a desert bird in the far southwest United States and Mexico, is a distinctive primary cavity nester. It excavates its nest hole in a giant, living saguaro cactus. As the cactus develops a layer of tissue around the 'wound', it forms a secure cavity nest. The woodpecker also digs nest holes in dead wood in cottonwood and willow trees.

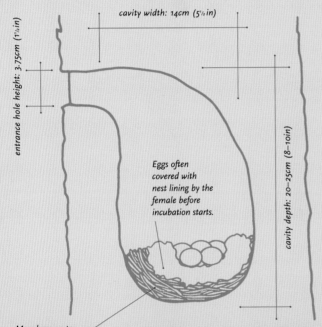

entrance hole height: 3.75cm (1½in)

cavity width: 14cm (5½in)

cavity depth: 20–25cm (8–10in)

Eggs often covered with nest lining by the female before incubation starts.

Moss base up to 5cm (2in) deep no matter the size of cavity, meaning that the requirement for moss is sometimes huge. The cup is positioned on the moss bed on the opposite side to the entrance hole.

FIG. 2
FOUND TREE HOLE
A secondary cavity nester, the Great Tit (Parus major) of Eurasia takes over an existing tree hole. Within the hole, a bed of plentiful moss and a little grass is constructed to support a cup thickly lined with hair, wool and feathers felted together. The cup is always on the side farthest from the entrance. The female builds alone, with construction time varying enormously from 2 to 20 days.

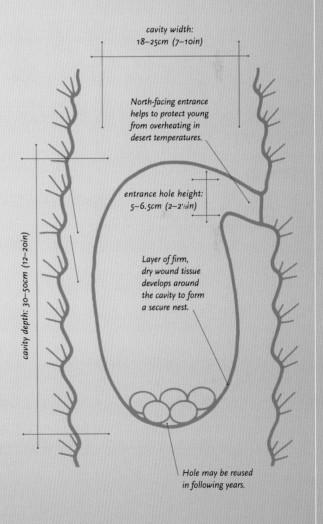

cavity width: 18–25cm (7–10in)

North-facing entrance helps to protect young from overheating in desert temperatures.

entrance hole height: 5–6.5cm (2–2½in)

Layer of firm, dry wound tissue develops around the cavity to form a secure nest.

cavity depth: 30–50cm (12–20in)

Hole may be reused in following years.

FIG. 2 GREAT TIT NEST | FIG. 3 GILA WOODPECKER NEST

Narina Trogon Tree Hole

There are nearly 40 species of trogon, most of which are dazzlingly hued tropical birds of warm, forested lowlands. They favour tree cavities and rotten stumps for their nests, and are both primary and secondary cavity nesters, adopting old nest sites or engineering their own nests by biting or gouging out soft wood. The Narina Trogon (*Apaloderma narina*) is a widespread species from West to East Africa, then south to Cape Province. Like other trogons, it generally nests in tree cavities. Whereas many hole-nesting species have a nest hole that is just small enough for them to get through, the trogon prefers a wider entrance. When studying Narina Trogons, naturalist-photographer Hugh Chittenden found that they would not enter a well-placed nest box with a hole 5cm (2in) in diameter, the perceived ideal dimension for this size of hole nester. However, on hearing of a natural nest hole 15cm (6in) wide, he enlarged the nest-box entrance, which had the effect of attracting a breeding pair.

Insect nests

Although the main illustration shows a typical tree cavity nest, trogons are one of a number of cavity architects who also excavate holes inside arboreal termite nests. A mixture of rotting wood and faecal cement, the material of the termite nest is an ideal hardness for the bird to excavate. Cotingas, puffbirds, kingfishers and some parrots also use termite nests, while the Violaceous Trogon (*Trogon violaceus*) will make its home inside a wasp nest. The Peach-fronted Parakeet (*Aratinga aurea*) shown here is nesting in an arboreal termite nest in Mato Grosso do Sul State, Brazil.

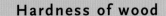

Hardness of wood

This male Narina Trogon is shown leaving the nest hole. The hole is usually 3–5m (10–16ft) above ground, is 20–60cm (8–24in) deep and is unlined. The nest site is inspected frequently by the pair before final acceptance by the female, and one of the key factors in choosing a site is the hardness of the wood. It has to be soft enough for the trogon to excavate with its beak, but hard enough to create an egg chamber that will not crumble. Old tree stumps and termite nests are ideal – trogons have been observed landing on fallen trunks and striking the wood with their tails, perhaps to test its firmness. A nest cavity may be used for several years.

Liquid defence

The large entrance hole to the nest makes it accessible to predators. The birds are constantly on guard: the male incubates for much of the day, and the female at night, with changeovers early morning and late afternoon. The trogons can deter predators by secreting an evil-smelling liquid from their preen gland, which is hard to wash away. Here, a female trogon defends the nest entrance.

CASE STUDY
Great Hornbill

Great Hornbills are huge birds that nest in large hollow spaces in trees in the jungles of India, the Himalayas, Indochina and Malaysia. They are particularly large for cavity-nesting species. They do not excavate; instead, a pair searches the forest for a suitable cavity. Once found, the female customizes the entrance before ensconcing herself for the entire period of incubation and chick rearing.

Classification

ORDER	Bucerotiformes
FAMILY	Bucerotidae
SPECIES	*Buceros bicornis*
RELATED SPECIES	Kingfishers, bee-eaters
NEST TYPE	Tree cavity
SPECIES WITH SIMILAR NESTS	Woodpeckers
NEST SPECIALIZATION	Female sealed within

Nest defence

From the outside, the female hornbill uses mud to seal cracks and to narrow the entrance aperture. She then squeezes into the cavity and uses her own faecal matter to seal the entrance until it is a narrow vertical slit. Sealing the nest cavity secures it from mammalian nest predators, such as yellow-throated martens and binturongs. Using her huge bill, the female can kill or ward off any creature small enough to slip through the slot entrance.

Maternal confinement

The female hornbill remains in the cavity for months until she has produced young of nearly adult size. During this period, she will drop her wing and tail feathers, being flightless as new feathers grow. The female and the young she produces are completely dependent on the male for food. The fruit he passes through the slot also provides all the water that the occupants require. When she finally leaves the nest, the young chicks rebuild the entrance.

Sanitation and temperature control

Both sanitation and temperature can cause potentially grave problems for a female hornbill and her young trapped within the chamber. To ensure that the nest stays clean and hygienic, both the female and the nestlings squirt their droppings out of the slot opening. The slot entrance is also essential for controlling the temperature of the nearly sealed-off nest. Hot air from within the nest rises and escapes through the top of the slot, while cooler air from outside flows in at the bottom of the slot, creating a neat air-conditioning system.

LEFT
TARICTIC HORNBILL SEALED NEST
A female Taractic Hornbill (Penelopides panini) almost entirely seals herself in with a ring of mud, through which the male will pass her food. This species is endemic to the Philippines.

BELOW
ORIENTAL PIED HORNBILL BREAKOUT
This illustration is based on a photographic sequence of a female Oriental Pied Hornbill (Anthracoceros albirostris) of southern India, Sri Lanka, Malaysia and Indonesia. Having raised her young, she uses her powerful beak to break free of the confines of her nest.

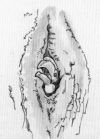 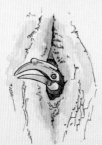    

1. The female breaks the mud seal.

2. She breaks down the entrance hole.

3. The hole is gradually widened.

4. The head and top of the body emerge.

5. The female's wing pulls free.

6. The female is free to fly the nest.

CASE STUDY
European Bee-Eater

The European Bee-eater is a primary cavity nester with special techniques for ventilating the tunnel it builds. A summer visitor to warm latitudes in the Northern Hemisphere, from Spain and North Africa eastward to Pakistan and Afghanistan, it breeds in open habitats such as grasslands. Bee-eaters breed in colonies, and breeding pairs are aided in chick-rearing by male 'helpers'.

Nest and habitat

This species excavates a tunnel in a sandy or earth bank bordering a river. Both sexes dig the entrance tunnel. In flat ground, a short vertical hole leads to a horizontal tunnel; in a bank, the tunnel is horizontal or slopes up from the entrance. The entrance hole is 10–13cm (4–5in) in diameter, and the tunnel has a diameter of 6.5–7.5cm (2½–3in). The tunnel's average length is 120cm (4ft), ending in the wider and higher nest chamber. A nest takes 2–3 weeks to complete.

Ventilation

As the birds enter and leave the narrow tunnel, their bodies press tightly against the walls. Their movements act like a piston, pushing in fresh air and expelling stale air. Although ammonia and carbon dioxide levels become high, this behaviour, combined with gusts of wind across the entrance hole, help to keep the nest ventilated.

Colony and breeding

Bee-eaters are gregarious all year round, migrating and wintering in flocks of hundreds. They breed in colonies from a few pairs to several hundred, and there tend to be more males in a colony. Typically, nest holes are a few metres apart; a colony in France, for example, featured 102 nests in 1km (0.6 miles) of riverbank. In tropical species, this leads to 'helpers' attaching themselves to breeding pairs; 'trios' have been observed in the European species, too. Both sexes of the pair incubate the eggs, and they and the helper feed the chicks. From a clutch of six or seven white eggs, incubated for 20 days, young will fledge after another 20–30 days.

Classification

ORDER	Coraciiformes
FAMILY	Meropidae
SPECIES	*Merops apiaster*
RELATED SPECIES	Kingfishers, rollers, todies, motmots, hoopoes
NEST TYPE	Excavated tunnel
SPECIES WITH SIMILAR NESTS	Kingfishers, jacamars
NEST SPECIALIZATION	Ventilated tunnel

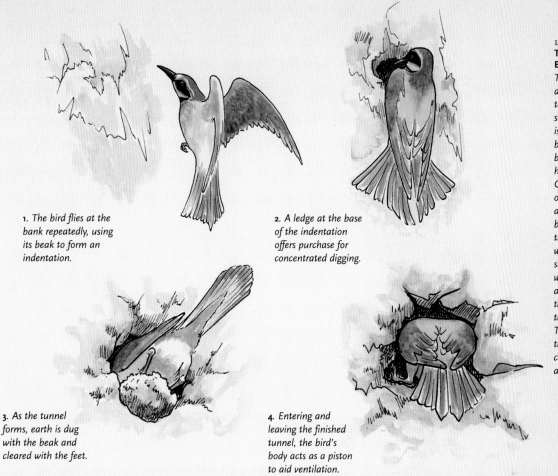

1. The bird flies at the bank repeatedly, using its beak to form an indentation.

2. A ledge at the base of the indentation offers purchase for concentrated digging.

3. As the tunnel forms, earth is dug with the beak and cleared with the feet.

4. Entering and leaving the finished tunnel, the bird's body acts as a piston to aid ventilation.

LEFT
TUNNEL EXCAVATION
The first act in digging a tunnel is to fly at the bank's surface until a hollow is formed (1.). The birds can then perch by hanging onto the hole's bottom lip (2.). Once enough sand or soil has been dug away to enable the birds to get into the hollow, they dig with their beaks and shovel the spoil out with their feet (3.), and continue in this way until the tunnel is complete. The movement of the birds in the completed tunnel aids ventilation (4.).

LEFT
COLONY BEHAVIOUR
Bee-eaters, unlike some species, are often comfortable with having near neighbours. Their feeding, preening, bathing and defence of the colony are infectious behaviours that soon bring more birds together. A group such as this, on a favourite perch, is ideally placed to catch an insect or attack an intruder. However, when each has a nest, they can be aggressive towards their neighbours.

CASE STUDY
Burrowing Owl

The Burrowing Owl can build, but commonly finds excellent defensive architecture ready made. Its bold nature is suited to aggressive adoption of existing burrows, often those of prairie dogs. Breeding from southwest Canada south to the western United States and Tierra del Fuego in Argentina, the owl's habitats include deserts, prairies, agricultural areas and grasslands, including golf courses and airfields.

Nest and habitat

A secondary cavity nester, the Burrowing Owl's capacity to take over the ground burrow of a mammal, especially the prairie dog, affords it an instant nest with established defences against predators. Such burrows may be several feet long, and are found in a range of desert, prairie and grassland habitats. If the ground is soft, Burrowing Owls are also proficient enough engineers to dig their own tunnels. The nest chamber is lined with a variety of materials, most interestingly animal dung, which may be placed at the burrow's entrance. Research suggests that the dung not only helps to control the tunnel's microclimate, but also attracts the insects that form the owl's diet.

Satellite burrows

Satellite burrows are often located near to the main nest. These are used as roosting sites by the adult male during nesting and by the fledglings later. The same burrow or a nearby one can be reused the following year.

Artificial nests

Although some South American populations are increasing, even inhabiting city parks or newly deforested regions, North American birds are threatened by loss of habitat and the pest control of prairie dogs. Conservation measures include artificial nest sites made of plastic tubing and nearby perches; some successful pairs have been enticed to move of their own accord to alternate breeding grounds, away from proposed building developments.

Classification

ORDER	Strigiformes
FAMILY	Strigidae
SPECIES	*Athene cunicularia*
RELATED SPECIES	Little Owl, Spotted Little Owl, Forest Spotted Owl
NEST TYPE	An existing hole
SPECIES WITH SIMILAR NESTS	Chickadees, Purple Martin, starlings
NEST SPECIALIZATION	Adopted prairie dog burrows

CASE STUDY

Red-Cockaded Woodpecker

The Red-cockaded Woodpecker is the only woodpecker to excavate nest holes exclusively in living pine trees, and it uses tree sap to adorn its defensive architecture. Resident in the southeast of the United States in mature pine forest, its habitat has been much reduced. The woodpecker and its nest play a vital part in the region's biodiversity, other birds, animals and insects using its abandoned nest holes.

Nest and habitat

These woodpeckers breed only in the living trees of old pine forests that are 60–80 years old. The nest hole is located 4–18m (12–60ft) high, has an entrance 5cm (2in) in diameter and is 20–30cm (8–12in) deep. The trees are often those with Red-heart Rot fungus disease, which softens the heartwood. Both birds of the breeding pair drill. The hole chosen for a nest cavity may take 1–3 years to finish and be in use for over 20 years.

Nest defence

Red-cockaded Woodpeckers defend their nest hole with a unique building activity. They pick off a wide area of bark for several centimetres all around the entrance hole. The sticky sap exudes and becomes a 'snake trap' – a defence against the large Black Rat Snake (*Elaphe obsoleta*), which is an expert tree climber and potential predator. The Longleaf Pines (*Pinus palustris*) produce the best flow and the birds keep the flow going. The smears of sap are conspicuous over a wide area around the nest's entrance.

Breeding and clan

Red-cockaded Woodpeckers live in extended family groups, consisting of a breeding pair, young birds and sons of the breeding male, one of which inherits the territory. Today, there are around 5,000 groups divided into about 30 significant populations from Florida to Virginia to Texas. All members of the clan help to defend the territory, incubate the eggs and feed the young. Juvenile females usually leave the group to find a solitary male.

Classification

ORDER	Piciformes
FAMILY	Picidae
SPECIES	*Picoides borealis*
RELATED SPECIES	Toucans, barbets, hornbills
NEST TYPE	Self-excavated hole
SPECIES WITH SIMILAR NESTS	Other woodpeckers, kingfishers
NEST SPECIALIZATION	Excavated holes in living wood

CHAPTER THREE

Platform Nests

In the largest examples, platform nests can form some of the most enduring architectural structures, but they range in size from only a few centimetres across to huge piles some metres in diameter. Protection from predators is also provided in different ways, from the great height and panoramic view of the nest site to the platform being completely surrounded by water. The nature of the materials used to secure the platform nest is closely related to the size of the architect and its capacity to lift and assemble the building blocks.

Platform nests are found in different habitats throughout the bird world. Some birds build their platform in, and indeed from, reed beds. Others nest in bushes or trees and build with dead sticks and fresh twigs. Both sexes collect and build, gradually adding materials to their pile and using different techniques to make the nest secure. These range from artful picking and dropping to simple weaving.

The characteristic technique of the platform builder is 'piling up', a term coined by the zoologist Mike Hansell. At its simplest, one twig or branch is dropped onto another. However, interlocking can occur, either because of the structure of the touching twigs, or through the rudimentary attempts shown by some species to shake, bend or twist twigs to make them secure. Larger nests depend upon gravity and the simple weight of material to make the nest secure; they lack a deep cup to hold the eggs but the size of nest relative to the eggs compensates. A shallow cup in the centre may be lined with finer material such as leaves.

Many birds of prey build by piling up sticks and twigs. Some, including the Swallow-tailed Kite (*Elanoides forficatus*) of North America and the related Black-winged Kite (*Elanus caeruleus*) of Europe, create thin, flat structures. Several species, such as the Osprey (*Pandion haliaetus*), build up vast nests over years in a tree or on a specially erected platform on a pole, repairing them as necessary. The nests can grow to 90cm–1.5m (3–5ft) across.

Platform nests are often built on a conspicuous site to provide a good lookout point. One of the most impressive belongs to the Bald Eagle (*Haliaeetus leucocephalus*), whose eyrie is highly visible in a tree or on a rocky ledge. In contrast, species such as the Eurasian Bittern (*Botaurus stellaris*) and Magpie Goose (*Anseranas semipalmata*) build a reed platform at ground level.

Some birds utilize man-made structures for their nests. The Black-legged Kittiwake (*Rissa tridactyla*) often locates its nest on a building, while the White Stork (*Ciconia ciconia* – see photograph above – nests atop chimneys and utility poles. Haverschmidt (1949) dates one remarkable White Stork nest back as far as 1549.

RIGHT
BALD EAGLE PLATFORM
An adult Bald Eagle at its treetop eyrie in Florida. This nest is a good example of a conspicuous lookout.

BLUEPRINTS

Platform Nest Structures

The simplicity of the architectural blueprint for the piled-up platform nest is deceptive, as it creates some of the most monumental and enduring structures in the avian world. Birds of prey including eagles, kites and ospreys build platforms, as do herons, egrets, storks and spoonbills. The Common Wood Pigeon (*Columba palumbus*) makes a rough lattice pattern, while the Magpie Goose (*Anseranas semipalmata*) creates a well-matted platform.

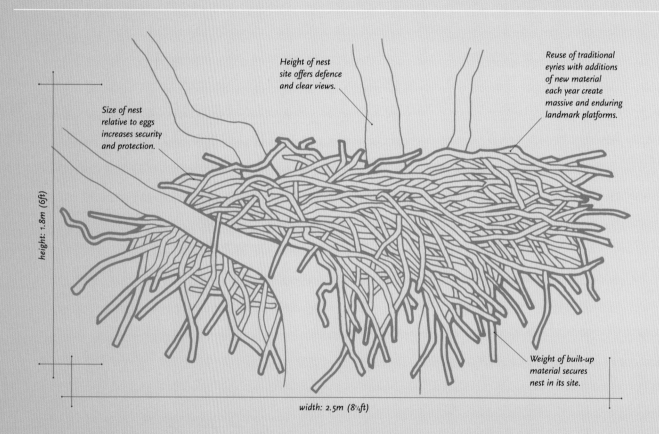

Height of nest site offers defence and clear views.

Reuse of traditional eyries with additions of new material each year create massive and enduring landmark platforms.

Size of nest relative to eggs increases security and protection.

height: 1.8m (6ft)

Weight of built-up material secures nest in its site.

width: 2.5m (8¼ft)

FIG. I
HIGH, HEAVY PLATFORM
The Bald Eagle (Haliaeetus leucocephalus) commonly forms its nest in a tree or on a cliff. The platform has good visibility and space for a clear takeoff, and is just a few hundred metres from water. In the north of its range, the nest site may be 24m (80ft) or more high, but it is no more than 6m (20ft) *high in the Florida mangrove swamps. In the treeless Aleutian Islands, Bald Eagles nest on the ground. Small sticks are picked up or broken off and carried in the beak; large ones are* *carried in the talons. In the centre of the platform a cup 7.5–13cm (3–5in) deep, lined with grass and mosses, is formed. Large nests can weigh two tonnes and last for over 50 years.*

FIG. I BALD EAGLE NEST

VARIETIES OF STRUCTURE

Enormous platform nests are built at dizzying heights
in trees or on cliffs using branches, sticks and twigs.
Platforms are also constructed in water using reeds.
Structures may be lined with grass, mosses, fine twigs,
leaves, wool and artificial materials. The simplest platforms
are unbound but some birds, such as the White Stork
(*Ciconia ciconia*), utilize a cement of earth, turf and dung.

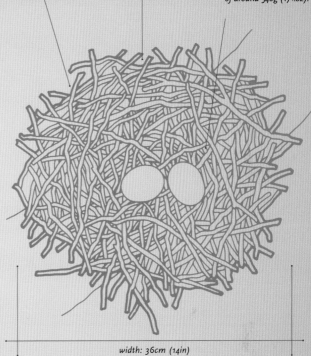

A flat, level or spreading branch provides the site for a flimsy platform.

Construction is in a lattice pattern of 200 piled-up twigs.

A lining is formed from thinner twigs weighing just a few grams each, less than 1.5 per cent of the bird's body weight of around 540g (17½oz).

width: 36cm (14in)

FIG. 2
PILED LATTICE PLATFORM

The platform of the Common Wood Pigeon is constructed from about 200 twigs up to 20cm (8in) long piled one on top of another in a rough lattice pattern. There is a lining of thinner but not necessarily shorter twigs. Early in the season, nests are commonly in ivy or evergreen trees, while later broods are often hidden in the leafy shrub layer of hazel, elder, hawthorn or honeysuckle trees.

Additional vegetation is added until the young are well grown.

On a site of matted stems and roots at water level, dead reeds are piled up to form a flat and circular platform.

average height: 38cm (15in)

average width: 61cm (2ft)

FIG. 3
REED-BED PLATFORM

The nest of the Eurasian Bittern (Botaurus stellaris) is built among the previous year's reeds. The female Bittern collects a loose heap of dead reeds, mostly from nearby, until it forms a circular, rather flat platform resting on matted vegetation. It is lined with a little finer vegetation. The male attracts one or more females into his territory; whichever female he mates with builds the nest, incubates the eggs and rears the young.

FIG. 2 COMMON WOOD PIGEON NEST | FIG. 3 EURASIAN BITTERN NEST

MATERIALS AND FEATURES
Common Wood Pigeon Nest

Common Wood Pigeons (*Columba palumbus*) nest throughout wooded Europe, building rickety nests from twigs in the fork of a tree (often a horizontal one). Evergreens or ivy provide sites early in the season; leafy trees, such as hawthorn, blackthorn and elder, later. The nest is on average 3–5m (9¾–16¼ft) above the ground. An English nest in a hawthorn from which the two young had just fledged measured 25 x 36cm (10 x 14in). It was faintly concave, without a definite cup, but the centre of the dip had finer twigs as a slightly softer lining. Almost all the twigs were of Silver Birch, which was the tree whose canopy overhung the hawthorn. Many of the twigs had multiple branches and were locked together. Outer twigs were bent around and pushed into the platform and the whole was cemented together with the youngs' droppings. At this stage it weighed 255g (9oz). A new nest commonly weighs closer to 150g (5¼oz).

Precise materials

The behaviour of the male suggests that precise selection of materials is essential. Twigs are mostly brought to the nest site by the male. When searching for these on the ground, he picks up and drops twigs repeatedly before making a choice. He also expends much effort twisting and tugging selected twigs to break them off the tree. The male brings the twigs in his bill to the female. She does the building, laying out the twigs one on the other, 'beam' upon 'beam', as she turns around.

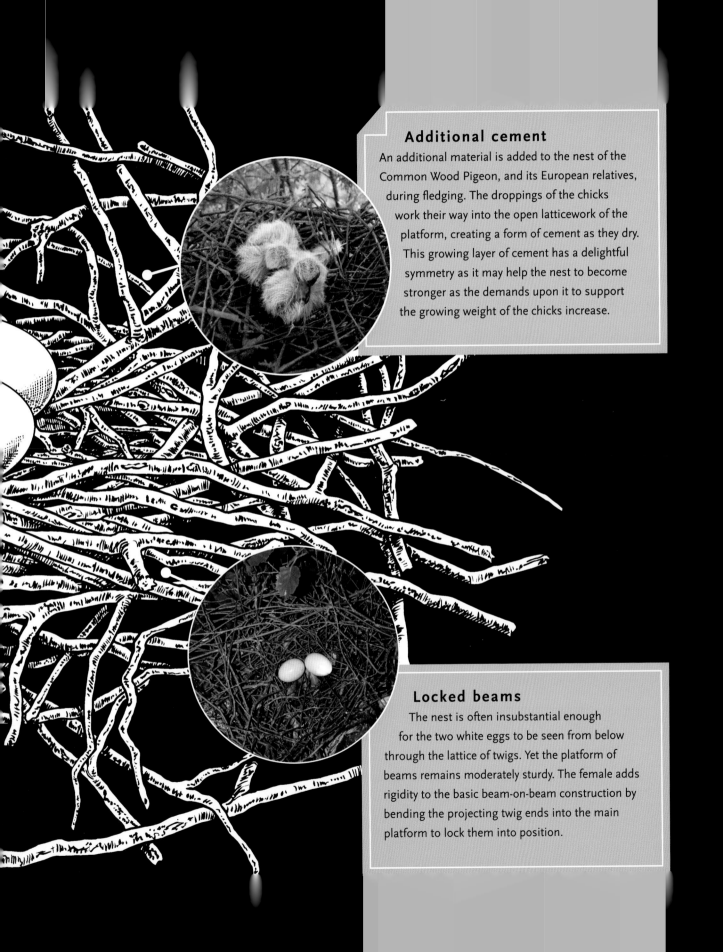

Additional cement

An additional material is added to the nest of the Common Wood Pigeon, and its European relatives, during fledging. The droppings of the chicks work their way into the open latticework of the platform, creating a form of cement as they dry. This growing layer of cement has a delightful symmetry as it may help the nest to become stronger as the demands upon it to support the growing weight of the chicks increase.

Locked beams

The nest is often insubstantial enough for the two white eggs to be seen from below through the lattice of twigs. Yet the platform of beams remains moderately sturdy. The female adds rigidity to the basic beam-on-beam construction by bending the projecting twig ends into the main platform to lock them into position.

Building a Reed Platform

The Magpie Goose (*Anseranas semipalmata*) is in a waterfowl family on its own, most closely related to screamers and whistling ducks. It is found in the swamps, mangrove flats, and floodplains of northern, tropical Australia. The birds breed just after the wet season, generally in March to May, but as early as January or as late as June, depending on the rains. If the wet season fails, they will not breed. Generally, the nest site is near the middle of a swamp. Magpie Geese nest colonially. The male builds the nest using rush or reed stems to build a platform. Some males have two mates; 5–14 eggs may be laid, and all three birds help to raise the young.

1. The Magpie Goose nest site is in the middle of a swamp, and the male bird begins construction by bending down and piling up rooted rush stems.

2. The reed stems are clutched in the beak and folded down one at a time to construct a circular, open platform.

3. The Magpie Goose rotates as it engineers the circular platform.

LEFT
MAGPIE GOOSE
The Magpie Goose is a large, pied bird with a wingspan of up to 5⅓ft (162cm). Its long, yellow legs have only partially webbed feet. Unusually for a goose, it often perches in trees.

4. The rotation of the bird effectively weaves the reeds into a radial pattern to give the nest added strength.

6. The bird tramples down the reeds with its feet to build an increasingly bulky and solid platform.

7. The combination of architectural techniques creates a semi-floating platform supported by the standing plants.

8. The final platform is a trampled-down tussock of vegetation half floating in water up to 3ft (1m) deep. The nest has a deep cup at the top to hold the eggs.

5. Sometimes the male Magpie Goose will bend several stems at once by catching them in the crook of his neck.

ROTATION
The laying of the stems is engineered in a careful rotation, which results in a well-matted platform.

CASE STUDY
Golden Eagle

The Golden Eagle constructs a vast platform nest high in a tree or on the ledge of a cliff. It competes successfully for prime sites, and particular crags have been used by successive pairs for up to a century. Found across the Northern Hemisphere, from northern Siberia and Alaska south to the European Alps, Mexico and the Himalayas, the Golden Eagle nests in several sites within its territory.

Classification

ORDER	Falconiformes
FAMILY	Accipitridae
SPECIES	*Aquila chrysaetos*
RELATED SPECIES	Other *Aquila* eagles, hawks, vultures, harriers
NEST TYPE	Platform
SPECIES WITH SIMILAR NESTS	Herons, egrets, storks
NEST SPECIALIZATION	Huge nests on multiple sites

Nest and habitat

Eagles and other large raptors are unique in having a specific name for their nest – the eyrie. This is built either on the ledge of a cliff or on a tree, often at the crown. A pair of birds usually have two or three nests. Some pairs exclusively have crag nests, others always trees and some use both. Observers have recorded as many as 10 sites for a single pair.

Nest construction

Golden Eagles are known to live for about 20 years, and they pair for life. The pair share the nest building and may build at more than one site in a year. When first constructed, a cliff nest is little more than a scrape surrounded by a ring of sticks up to 4cm (1²⁄₃in) thick, with a lining of a little grass, heather, wood rush, wool and leaves. It is added to each consecutive year, until it reaches 1–3m (3⅓–10ft) in diameter and 2m (6½ft) tall. The hen probably does most of the building, picking up or breaking off sticks, pulling up grass and heather, and piling up the profuse material. Tree nests can become huge, as wide as 2.4–3m (8–10ft) across and 5.2m (17ft) deep. Two eggs are laid, but in 80 per cent of nests only one chick survives to fledging.

Nest adornment

As the breeding season goes by the birds add sprays of green vegetation predominantly from pine trees, but almost any greenery if pine is not available. The reason is disputed. It may be to introduce clean material; to employ insecticidal and antibacterial properties; or, as the late David Bannerman put it, to add 'a touch of beauty to the structure'.

LEFT
CRAG NEST
The powerful Golden Eagle can lay claim to prime cliff sites, which can provide nests for decades. The size of the cliff ledge dictates the nest's size. The nest may stretch along the ledge and have a cup formed in a different place each year. This crag nest demonstrates the good visibility afforded by a prime site, and the space around the eyrie which enables the adult with its 1.9–2.3m (6–7½ft) wingspan to take off easily.

LEFT
NEST ADORNMENT
Tree nests start as a shallow platform of sticks and have a lined cup, especially with wood rush, but also grass and heather. This adult standing with its chick is on a nest near Konny Dvor, 280km (175 miles) northeast of Minsk, Belarus. It shows the green pine branches that are often added to the platform.

CASE STUDY
White Stork

Man-made architecture commonly forms a towering foundation for the White Stork to construct its highly distinctive and attractive nest. Gregarious summer visitors to Europe, North Africa, the Middle East and southwest Asia, they winter in sub-Saharan Africa and India. White Storks construct extravagant, sprawling platforms that they reuse and extend each season.

Nest and habitat

In the wild, White Storks build nests on trees and cliff ledges, while in towns and villages they characteristically build on roofs, towers, ruins, chimneys, utility poles and man-made platforms erected on poles. A drive along a main road in Tunisia can be marked by nest after nest located on electricity pylons, built on the steel platforms at the very top.

Nest building

In the breeding season, the male White Stork usually arrives first and chooses a site or reclaims the nest from the previous year. He defends it vigorously and starts to build or repair it, assisted later by the female. The nest takes the form of a huge pile of dead branches and sticks up to 3–3.75cm (1⅓–1½in) thick. The materials are glued together with earth, turf and dung, all collected and carried with the bill from as far as a kilometre (½ mile) away. The central depression is lined with twigs, grass and sometimes artificial materials. A new nest can be finished in eight days and measure 80–150cm (2⅔–5ft) across by 90–180cm (3–6ft) deep, but one used over many years may become nearly twice as big.

Associated nests

Older storks' nests are so deep that Spanish Sparrows (*Passer hispaniolensis*) or House Sparrows (*Passer domesticus*) may add their own untidy, domed nests of grass lined with feathers around the bottom. The chirping small birds do not seem to trouble the rightful owners of the main nest, although they can be very aggressive towards other storks that come too close.

Classification

ORDER	Ciconiiformes
FAMILY	Ciconiidae
SPECIES	*Ciconia ciconia*
RELATED SPECIES	Bitterns, herons, egrets, ibises, spoonbills
NEST TYPE	Stick platform
SPECIES WITH SIMILAR NESTS	Relatives except bitterns
NEST SPECIALIZATION	Huge nest on man-made structure

TOWERING PLATFORM
This White Stork nest is at the top of the tower of the necropolis of Chellah, at Rabat, Morocco. It has clearly been used for several years, as its size testifies. The size of the nest and the stork can be measured by the sparrow perched on the right, whose own nest may be constructed at the bottom fringes of the stork's nest. The combination of man-made and avian architecture creates one of the great spectacles of the natural world.

CHAPTER FOUR

Aquatic Nests

Specialist architects are required to build on water, which places unique demands on the nest-building techniques while also offering immediate protection from land-based predators. Only four bird families construct truly aquatic nests: jacanas, marsh terns, grebes and rails.

These architects build using aquatic plant materials that generally contain air spaces. When detached from the base, the plants tend to float – even the rotting stems and leaves. Two building styles are employed. Substantial mounds are piled up and either rest on the bed in shallows or form floating islands in deeper water. Lightweight rafts of plant material offer a less sturdy floating structure. In both cases, surrounding plants offer convenient anchors.

Aquatic nests have several advantages. The water acts as a protective moat and the nests may be well hidden; for example, grebes breed by lakes, pools or calm streams, bordered by vegetation that conceals the nest. A floating nest can also rise and fall with changing water levels. However, aquatic nests can be destroyed in bad weather or left high and dry by drought. An incubating coot may sometimes knock eggs out of the shallow cup of its nest. When an incubating jacana stands up, the movement often thrusts the eggs out of the flimsy nest and into the water.

To minimize their vulnerability, the young in aquatic nests hatch in a highly developed state and the breeding pair are very protective. The chicks become independent quickly and they leave the nest early.

The eight species of jacanas, or 'lily trotters', have unusually long toes and claws that enable them to walk across floating vegetation. And that is the form their flimsy nests take. That of the Pheasant-tailed Jacana (*Hydrophasianus chirurgus*) was described in 1952 by the naturalist W. W. A. Phillips as: 'A few floating waterweeds piled together to form a small blob almost awash'.

Marsh terns, found breeding as summer migrants in marshes and swampy grasslands by areas of open water, construct floating nests of varying degrees of strength, and also adopt the old nests of birds and mammals. Grebes can engineer more substantial structures. Clumsy on land, they need to swim right up to the nest. They dive to pick up rotting vegetation, which is then either built up from the bed of shallow water or anchored to surrounding reeds. The nest is trampled at the top to form a shallow cup.

The nest of the Common Coot (*Fulica atra*), a widespread European rail species, is also a substantial aquatic structure piled up in the shelter of reeds, rushes or cattails. Ornithologist Howard Saunders in 1899 recorded a nest firm enough to support a seated man who was up to his knees in water.

RIGHT
PIED-BILLED GREBE NEST
A Pied-billed Grebe (Podilymbus podiceps) incubates its eggs on an aquatic nest in Texas.

Aquatic Nest Structures

Aquatic nest blueprints take two main forms, the mound piled up from the bed of the site, and the true floating nest. A key engineering technique is the anchoring of the nest material to submerged vegetation. Aquatic structures are found in four families: the gallinules, rails and coots (Rallidae, about 140 species); the grebes (Podicipedidae, around 20 species); the jacanas (Jacanidae, 8 species); and the marsh terns (Sternidae, 3 species).

FIG. I
**PILED-UP
MOUND**
A bulky, cup-shaped mound of dead leaves and waterside plants forms the Common Coot's (Fulica atra) nest. It is built among growing vegetation in or by fresh water, and the material is piled up, mainly by the female. The male builds extra platforms for the young to rest on.

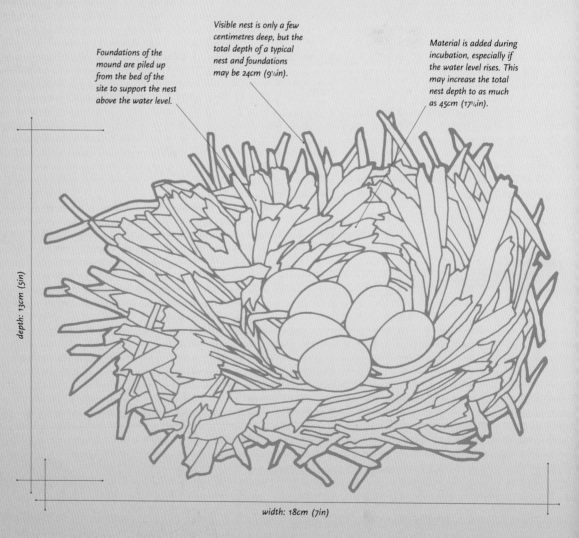

Foundations of the mound are piled up from the bed of the site to support the nest above the water level.

Visible nest is only a few centimetres deep, but the total depth of a typical nest and foundations may be 24cm (9¼in).

Material is added during incubation, especially if the water level rises. This may increase the total nest depth to as much as 45cm (17¼in).

depth: 13cm (5in)

width: 18cm (7in)

FIG. I COMMON COOT NEST

VARIETIES OF STRUCTURE

Substantial aquatic structures are built by the grebes, and the ornithologist K. E. L. Simmons, in his study of Great Crested Grebes (*Podiceps cristatus*) wrote that grebes are 'efficient if not very elaborate nest builders'. The Rallidae, who also build substantially, have varied nesting habits. Some produce covered nests of grass, some nest away from water. For example, the Common Moorhen (*Gallinula*

chloropus) builds a platform of mostly dead water plants. It usually constructs its nest among aquatic vegetation, but will often build it in trees or bushes overhanging or near water, even as high as 6m (2oft) or more in a willow. Jacanas build flimsy, seemingly unsafe nests, as shown on page 54, and the parents have to be vociferous defenders of the nests and their young.

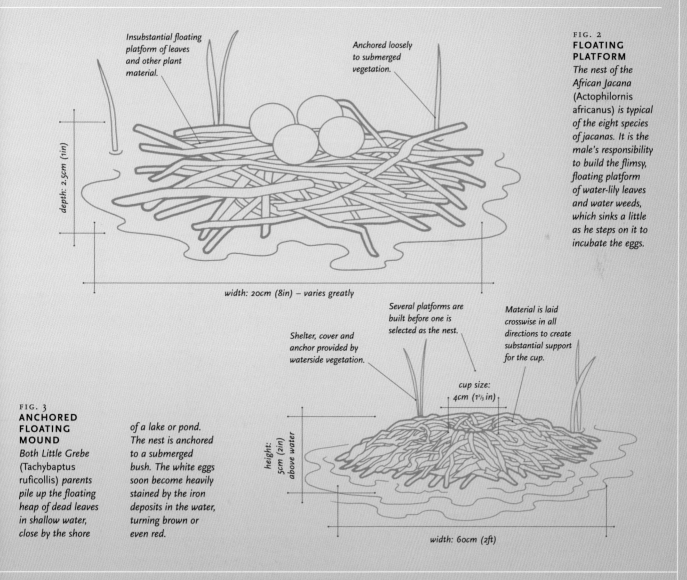

Insubstantial floating platform of leaves and other plant material.

Anchored loosely to submerged vegetation.

depth: 2.5cm ('1in)

width: 2ocm (8in) – varies greatly

FIG. 2
FLOATING PLATFORM
The nest of the African Jacana (Actophilornis africanus) is typical of the eight species of jacanas. It is the male's responsibility to build the flimsy, floating platform of water-lily leaves and water weeds, which sinks a little as he steps on it to incubate the eggs.

Several platforms are built before one is selected as the nest.

Material is laid crosswise in all directions to create substantial support for the cup.

Shelter, cover and anchor provided by waterside vegetation.

cup size: 4cm (1²/₃ in)

height: 5cm (2in) above water

width: 6ocm (2ft)

FIG. 3
ANCHORED FLOATING MOUND
Both Little Grebe (Tachybaptus ruficollis) parents pile up the floating heap of dead leaves in shallow water, close by the shore *of a lake or pond. The nest is anchored to a submerged bush. The white eggs soon become heavily stained by the iron deposits in the water, turning brown or even red.*

Fɪɢ. 2 ᴀғʀɪᴄᴀɴ ᴊᴀᴄᴀɴᴀ ɴᴇꜱᴛ | Fɪɢ. 3 ʟɪᴛᴛʟᴇ ɢʀᴇʙᴇ ɴᴇꜱᴛ

Great Crested Grebe Nest

The Great Crested Grebe (*Podiceps cristatus*) breeds widely on freshwater from the British Isles to China, in scattered populations in Africa, and in Australia. The pair builds one or more platforms by diving for weed and then piling it up until there is 'land' expansive enough to support the weight of both birds. The grebe usually nests in a separate, carefully anchored construction in the shelter of waterside vegetation. Among Great Crested Grebes, the male carries most of the sodden, decayed waterweed to the site, transporting it by the beakload. In shallower waters, the nest may reach to the bed, but otherwise it will be floating and anchored to surrounding vegetation. Most of the nest lies below the waterline – sometimes even the cup – making it an extremely damp home, which the chicks leave as soon as they have all hatched and are dry.

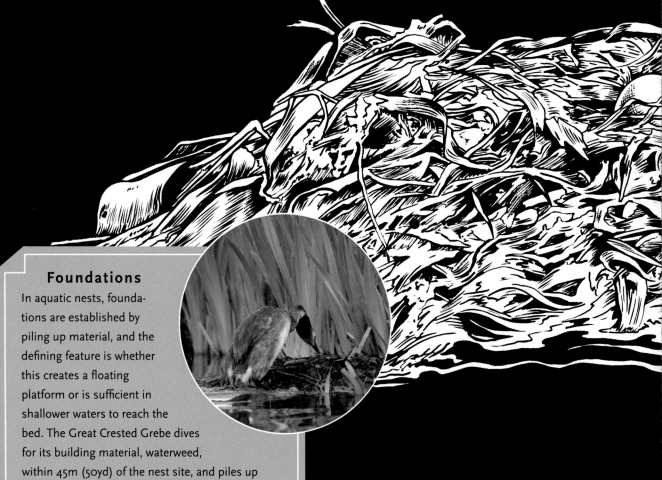

Foundations
In aquatic nests, foundations are established by piling up material, and the defining feature is whether this creates a floating platform or is sufficient in shallower waters to reach the bed. The Great Crested Grebe dives for its building material, waterweed, within 45m (50yd) of the nest site, and piles up profuse amounts. Observers recorded one pair transporting 100 beakfuls in less than an hour.

Elasticity

In severe weather the nest has to survive the motion of the water and the threat of flooding from the waves. This is where the elasticity of the building material and the capacity for an anchored floating nest to 'ride' the waves like a boat provides advantage. Rankin (1947) observed of one Great Crested Grebe nest in bad weather: 'One admired the skill of its builders in providing it with a certain amount of elasticity while at the same time keeping it firmly anchored.'

Mobile home

The nest and the parents are admirably adapted to their environment. As the grebe leaves the nest, it often covers the eggs by pulling weed and leaves over them. After hatching, the young are carried on the parents' backs for up to the first three weeks, providing a safe mobile home for fishing trips. Before the adult dives for food it shakes off the chicks.

CASE STUDY
Horned Grebe

Clumsy on land, the Horned Grebe constructs both floating and piled-up nests at the heart of its aquatic habitat. The bird winters as far south as Florida, the Mediterranean and Japan. It nests across the northern temperate zone of the Old and New Worlds, breeding on inland waters. The parents protect the young when they are very small, but the chicks are completely independent in about six weeks.

Nest and habitat

In the breeding season, the Horned Grebe prefers shallow lakes and ponds with fairly rich surrounding vegetation and a good food supply of small fish and other water creatures. Both sexes choose the nest site in new-growing sedge, horsetail or willow, in water up to 90cm (3ft) deep. The site is usually among enough vegetation to reduce the water's wave action, but within a few feet of open water.

Nest building

Courtship platforms are sometimes built before the nest. Although the pair chooses the nest site together, it is the male that usually starts building. The nest is built up from the bed of the pool or anchored to surface vegetation. Of 113 nests that were studied in Iceland, 97 were floating but anchored. Nest material is dived for and brought up in the bill. The pile may take only 3–4 hours to form; whatever the length of time, it is ready a few days before the first egg is laid. Both birds use their bills, feet and bellies to engineer a cup for the 4–5 chalky white eggs.

Breeding and defence

The chicks are fed with aquatic food bill-to-bill. While they are small, they climb unaided onto a parent's back for safety. They can dive for safety when 10 days old, by which time they are no longer welcome on board. A wing flap or a shake of the body by the adult sheds the young. During the weeks that the chicks are still dependent, they are defended vigorously by the parents with aggressive attacking behaviour and a distraction display of clumsy diving with big splashes.

Classification

ORDER	Podicipediformes
FAMILY	Podicipedidae
SPECIES	*Podiceps auritus*
RELATED SPECIES	22 other grebes
NEST TYPE	Aquatic
SPECIES WITH SIMILAR NESTS	Coots, moorhens, marsh terns
NEST SPECIALIZATION	Anchored floating nest

1. The grebe swims at speed up to the unobstructed nest.

LEFT
ACCESSING THE NEST
All grebes have elongated bodies with their feet placed a long way back, making them very clumsy on land. The ideal nest, therefore, is one that the bird can swim right up to (1.). On reaching the nest, the bird has to launch itself out of the water (2.) before driving itself into the nest cup (3.).

2. The bird uses its momentum to launch itself out of the water.

3. It propels itself the short distance to the cup.

LEFT
MOBILE HOMES
Grebes are unusual in that the young are carried on a parent's back from the chick's second day of life, unlike most other waterfowl, such as ducks and geese. All grebe chicks are characterized by their beautiful, striped downy plumage, especially on the head, hints of which remain until the juvenile feathered plumage is complete.

CASE STUDY
African Jacana

The insubstantial aquatic architecture of the African Jacana befits a bird that can move so easily over floating vegetation that it is known as the 'lily trotter'. All the jacanas are tropical birds resident on freshwater ponds. The African Jacana is found widely across Africa from Senegal to Sudan, and south to South Africa. To survive such a fragile nest, the eggs are waterproof and the young are precocial.

Mating platforms

At the beginning of the breeding season just after the rains, the male defends a small territory and tosses together small piles of weed to construct floating 'mating platforms'. A noticeably larger female courts him, and after mating she lays four eggs on one of the platforms. This floating platform then becomes the nest.

Female dominance

The female then deserts the nest, eggs and the male, and mates with other males. She can produce up to 30 clutches of eggs each season, from either the same partner or various partners. The Wattled Jacana (*Jacana jacana*) of South America is known to have a harem of five males. It is believed female dominance has evolved to compensate for an extraordinarily high rate of egg loss. Biologist Stephen Emlen of Cornell University has described the female as 'an egg-making machine'.

Defence and young

Male jacanas incubate the eggs and care for the young. The African Jacana has been seen to lift the eggs with his wings until they are under his brood patch and off the wet nest. When the male has to leave the nest, the eggs survive thanks to a waterproof cuticle. The precarious position of eggs and young on this nest is well shown here in this image from Chobe National Park, Botswana. The young chicks feed themselves as soon as they are dry. The male guards them and his alarm call will cause them to dive underwater. If danger persists, he will remove them elsewhere.

Classification

ORDER	Charadriiformes
FAMILY	Jacanidae
SPECIES	*Actophilornis africanus*
RELATED SPECIES	Seven other jacanas, other waders
NEST TYPE	Aquatic
SPECIES WITH SIMILAR NESTS	Other jacanas
NEST SPECIALIZATION	Floating nest, waterproof eggs

CASE STUDY
Black Tern

Black Terns are one of the three 'marsh terns'. The Canadian and American breeders winter in South America; the European birds are summer visitors that migrate to winter in central and southern Africa. They breed on fresh or brackish still water, 90–180cm (3–6ft) deep, and in swamps. The nest varies from a simple depression in vegetation to a substantial structure, and they also adopt Muskrat nests.

Nest and habitat

Black Terns nest in colonies of between 15 and 20 pairs, with nests 90cm–9m (3–30ft) apart. The nest is sometimes made from floating strands of vegetation anchored to one or two growing stems of a water plant, or built up on floating herbage. The eggs rest in a depression in the matted vegetation, and many nests are of strong construction. The pair pull material together to build a substantial mound, and the nest is lined with finer stems and seed heads. The eggs can withstand becoming wet, although not for long periods of time.

Adopted nests

Sometimes – and records suggest more so in North America than in Europe – a drier site is chosen on the top of an old grebe, coot or Muskrat nest, or even on the ground. One American observer watched a female bring a beakful of weeds 14 times in half an hour to the top of the remains of an old Muskrat house. On this solid foundation, the nest is little more than a scrape lined with some strands of grass or sedge.

Defence and young

Black Terns are very vocal at the nest site. They aggressively attack intruders – humans, animals or birds of prey – sometimes even striking the unwelcome visitor, to the accompaniment of intense screaming calls. The young are fed by both parents and are brooded in the nest for most of the first week, but they do wander into nearby vegetation when a few days old. They fledge when about three weeks old and become independent soon afterwards.

Classification

ORDER	Charadriiformes
FAMILY	Sternidae
SPECIES	*Chlidonias niger*
RELATED SPECIES	Over 40 other terns
NEST TYPE	Floating platform
SPECIES WITH SIMILAR NESTS	Jacanas, grebes
NEST SPECIALIZATION	Adopted Muskrat house

CHAPTER FIVE

Cup-Shaped Nests

The cup-shaped nest provides the classic model of avian architecture, and the most widespread of all nest types. Protection for eggs and young is provided by a robust cup with insulated lining; strong support and anchoring notably with spider silk; and the site commonly being raised above ground level. In this relatively secure environment, newly hatched young are born helpless. The chicks are altricial (their parents need to nourish them) and nidicolous (they remain in the nest for a length of time).

Cup nest builders use a great variety of materials — roots, stems, twigs, leaves, lichen, flowers and fruit. More exceptionally some birds, such as the American Robin (*Turdus migratorius*), add mud (see Chapter 7, pages 84–93, for the species that use only mud). Cup nests may show up to four elements in their construction: (1.) attachment, (2.) structure, (3.) lining and (4.) decoration. The nest materials are bound through a combination of engineering techniques and the materials' natural shapes and textures — which help them to 'interlock', a term first used by Mike Hansell.

The power of flight facilitates the building of an intricate nest. Naturalist Walter Scheithauer described how he watched a Brown Inca (*Coeligena wilsoni*) hummingbird choose fine strands of hemp 10cm (4in) long, place them crosswise on a branch, and then pick up loose ends and fly clockwise around the branch until a small cushion was fashioned. Material was added and a cup formed through a sequence of skilled body, head, wing and bill movements.

The size and shape of cup-shaped nests are directly related to the builder — this is normally the hen, and she forms the shape of the cup by twisting, turning and pressing her breast against the bottom and sides. Cup-shaped nests range in dimensions from the tiny hummingbird's half-walnut size to that of the Northern Raven (*Corvus corax*), which averages 1.7m (5½ft) diameter and 70cm (28in) depth.

The technically advanced architecture of a typical cup nest, such as that of the American Robin shown above, provides several key benefits. The cup is sturdy and well built, offering a good level of protection for eggs and chicks. It is securely attached to the tree, with birds such as the Chaffinch (*Fringilla coelebs*) constructing strong, fine anchors from spider silk. Excellent insulation keeps the temperature high during incubation — the Carrion Crow (*Corvus corone*), for example, constructs a luxurious lining of feathers, down and wool. The cup form also helps to keep the growing chicks together. Less favourably, the nest is not immune from predation and parasitism.

RIGHT
AMERICAN GOLDFINCH CUP
A female American Goldfinch
(Spinus tristis) *feeds her young,*
whose eyes are still closed.

BLUEPRINTS
Cup Nest Structures

Although the classic cup shape of this architecture is instantly recognizable, the great variety of materials used around the world ensures there is no one standard blueprint for the construction and composition of the outer nest layer, nor for the cup's lining. The four defining elements are the attachment, structure, lining and decoration. Nests are built by several thousand species of passerines including jays, blackbirds, flycatchers, thrushes, crows, finches, warblers and sparrows; cups are also built by many hummingbirds.

FIG. I
WATERSIDE WOVEN CUP
The female of the Red-winged Blackbird (Agelaius phoeniceus) *pair builds the deep cup alone. It is secured tightly to uprights of sedges, rushes, cattails or waterside trees. Construction materials include roots, grass, dead leaves and mud, all firmly shaped, and then lined with fine, dry grass or similar material. The ideal site is the middle of a marsh, swamp or water meadow. Territories are defended by the males, who gather a harem of several females – they choose such sites as the most promising in which to rear a brood, as has been well described by American naturalist Calvin Simonds.*

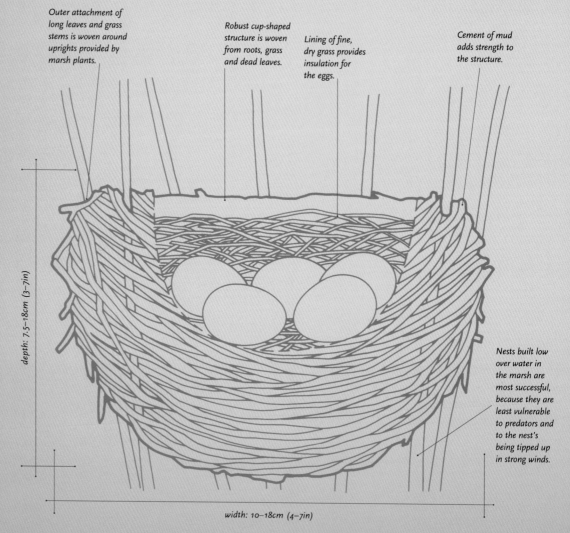

Outer attachment of long leaves and grass stems is woven around uprights provided by marsh plants.

Robust cup-shaped structure is woven from roots, grass and dead leaves.

Lining of fine, dry grass provides insulation for the eggs.

Cement of mud adds strength to the structure.

Nests built low over water in the marsh are most successful, because they are least vulnerable to predators and to the nest's being tipped up in strong winds.

depth: 7.5–18cm (3–7in)

width: 10–18cm (4–7in)

FIG. I RED-WINGED BLACKBIRD NEST

VARIETIES OF STRUCTURE

The size of the nest varies according to the bird – from a tiny hummingbird's nest to that of a large crow. Nests are built in a variety of ways – for example, by weaving leaves and grass around plant stems; by using dry grass stems piled into a cup shape; or by binding plant material with spider silk. They are lined with soft material, such as fine grass or hair. Cup nests may be suspended from marsh plants or branches overhanging water or attached to trees or bushes on land.

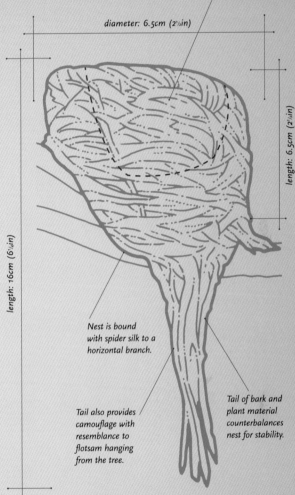

Precise, neat cup is usually placed less than 2.4m (8ft) above the ground or water.

diameter: 6.5cm (2½in)

length: 6.5cm (2½in)

Nest is bound with spider silk to a horizontal branch.

Tail also provides camouflage with resemblance to flotsam hanging from the tree.

Tail of bark and plant material counterbalances nest for stability.

length: 16cm (6¼in)

Stiff dry grass stems are bent at regular intervals to form a structure of short, hinged beams, which are stacked up into a cup shape.

An insulating lining uses finer grasses and hair – 180 horse hairs have been counted in the lining – and a firm woven rim often contains spider web.

cup depth: 4cm (1½in)

height: 6.5cm (2½in)

inner diameter: 6.5cm (2½in)

outer diameter: 10cm (4in)

FIG. 2
CUP OF HINGED BEAMS
This neat hinged-beam structure is made of dry grass stems, mainly by the female Eurasian Blackcap (Sylvia atricapilla). The male uses some grasses at the edge of the nest to bind the structure to the twigs of low bushes, honeysuckle or bramble. In spring, he makes up to seven simple nests or platforms, each taking 1–2 days to build. The bird employs song and display to lure a female to his work. She chooses a 'cock nest' and within 2–5 days completes the cup in which eggs will be laid.

FIG. 3
WINE GLASS
In Australia, the Grey Fantail (Rhipidura albiscapa) binds its nest to a horizontal branch that often overhangs water. The beautiful goblet shape is made of soft grasses, finely shredded bark and other plant material. The cup is an almost uniform buff colour, bound together and covered with spider web. The tail, or stem of the glass, may help disguise the nest to resemble flotsam caught in the tree, or may stabilize the cup perched on the branch. Such 'tails' are features of other flycatcher and some hummingbird nests.

FIG. 2 EURASIAN BLACKCAP NEST | FIG. 3 GREY FANTAIL NEST

MATERIALS AND FEATURES
Hummingbird Nest

Hummingbirds are tiny, compact birds that construct tiny, compact nests. A male Ruby-throated Hummingbird (*Archilochus colubris*) weighs $\frac{1}{10}$oz (3g) and a female $\frac{11}{100}$oz (3.3g).The nests of most take the form of simple cups, each smaller than a shot glass. Although some species suspend their nests, the typical hummingbird cup is attached directly on top of a tree branch. In most species, the female builds the nest, using small and light components that she binds together and adheres to the branch with spider or caterpillar silk taken from cocoons. Components such as bark, lichens, and leaf pieces are combined with the spider silk and molded into a cup nest. The interior is lined with softer material, such as plant fibers and animal hairs. Hummingbirds build among the most resilient nests of any bird. It can be argued that a nest that lasts so well has cost more effort than was needed. However, Ruby-throats and some other hummingbirds sometimes use an old nest as a base on which to build a new one the following season.

Insulating materials

Plant fibers, animal hair, and feathers are ideal materials to provide both a soft lining within the Ruby-throated Hummingbird's nest and insulation that ensures successful incubation of the eggs. The lining sits within a core structure built from bud scales and fluffy seed casings of thistle and dandelion, held in place with spider silk and occasionally pine resin, if available. The inward-curling lip of the nest is molded by the female. She pinches the combined materials between her bill, chin region, and chest, while rotating her body.

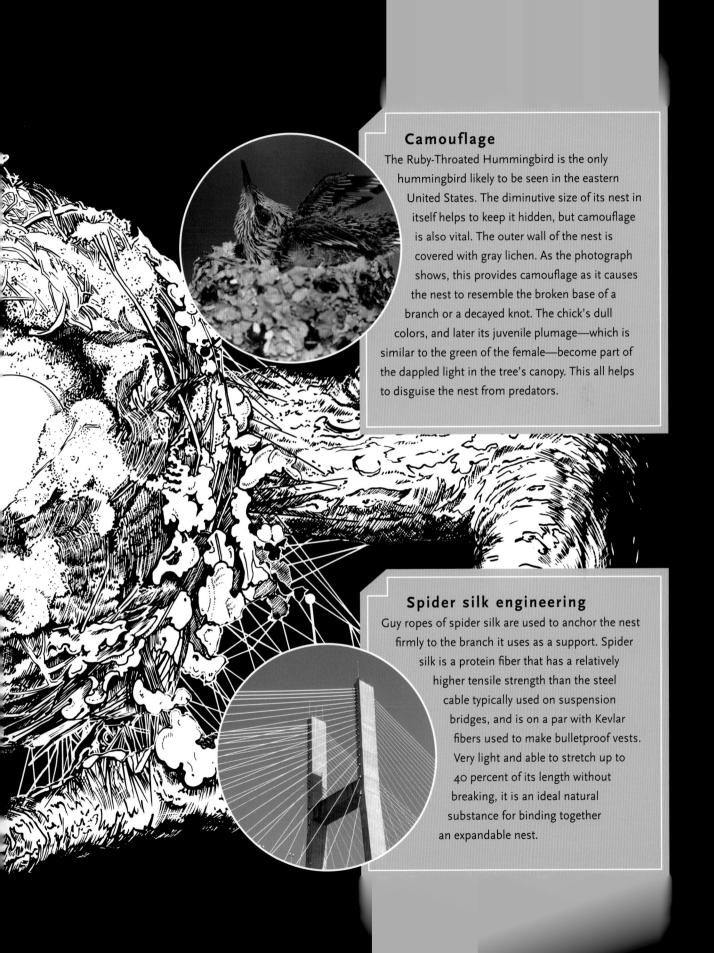

Camouflage

The Ruby-Throated Hummingbird is the only hummingbird likely to be seen in the eastern United States. The diminutive size of its nest in itself helps to keep it hidden, but camouflage is also vital. The outer wall of the nest is covered with gray lichen. As the photograph shows, this provides camouflage as it causes the nest to resemble the broken base of a branch or a decayed knot. The chick's dull colors, and later its juvenile plumage—which is similar to the green of the female—become part of the dappled light in the tree's canopy. This all helps to disguise the nest from predators.

Spider silk engineering

Guy ropes of spider silk are used to anchor the nest firmly to the branch it uses as a support. Spider silk is a protein fiber that has a relatively higher tensile strength than the steel cable typically used on suspension bridges, and is on a par with Kevlar fibers used to make bulletproof vests. Very light and able to stretch up to 40 percent of its length without breaking, it is an ideal natural substance for binding together an expandable nest.

Cup-Forming Techniques

The Song Thrush (*Turdus philomelos*) is the most widespread spotted thrush in Europe in woodland, parks and gardens. As with other members of the thrush family (see page 68), it builds a cup nest of grass and moss on a twig base. The cup lining is strengthened with mud or rotten wood but, uniquely among its relatives, it does not then add a soft inner lining of fine grasses. Such a hard, dry lining is very unusual and its function is undetermined. There are many stages in the construction of this intricate nest.

1. *The Song Thrush selects a site, typically a triple fork of laurel, on which to start building the platform attachment. The female works alone on the construction.*

2. *She constructs the platform from 'beams' of silver birch twigs which are up to 30cm (12in) long and 2–3mm (1/16in) thick.*

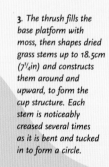

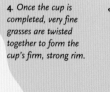

3. *The thrush fills the base platform with moss, then shapes dried grass stems up to 18.5cm (7¼in) and constructs them around and upward, to form the cup structure. Each stem is noticeably creased several times as it is bent and tucked in to form a circle.*

4. *Once the cup is completed, very fine grasses are twisted together to form the cup's firm, strong rim.*

LEFT
NEST SITE
This Song Thrush is feeding chicks in a nest that clearly shows its typical position – secured in a shrub's fork, and surrounded by thick cover.

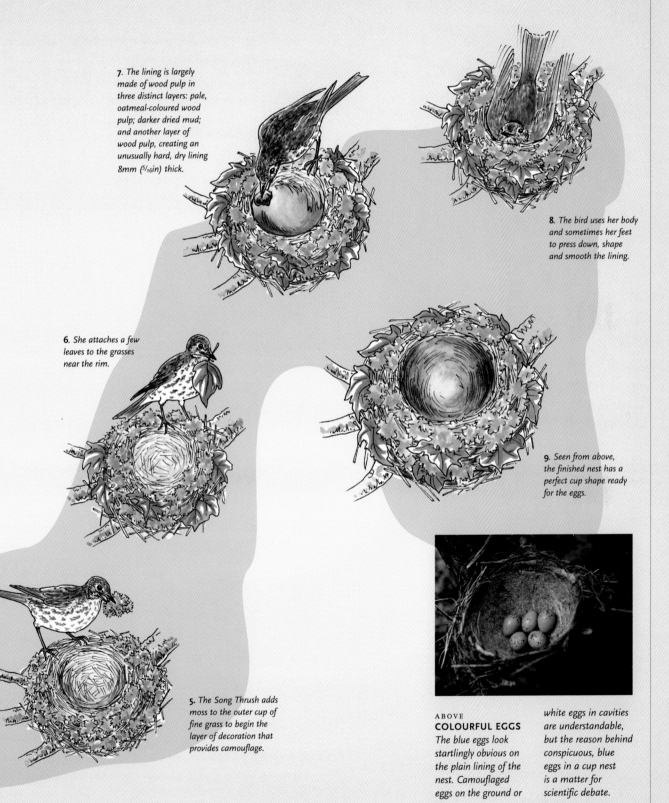

7. The lining is largely made of wood pulp in three distinct layers: pale, oatmeal-coloured wood pulp; darker dried mud; and another layer of wood pulp, creating an unusually hard, dry lining 8mm (⁵⁄₁₆in) thick.

8. The bird uses her body and sometimes her feet to press down, shape and smooth the lining.

6. She attaches a few leaves to the grasses near the rim.

9. Seen from above, the finished nest has a perfect cup shape ready for the eggs.

5. The Song Thrush adds moss to the outer cup of fine grass to begin the layer of decoration that provides camouflage.

ABOVE
COLOURFUL EGGS
The blue eggs look startlingly obvious on the plain lining of the nest. Camouflaged eggs on the ground or white eggs in cavities are understandable, but the reason behind conspicuous, blue eggs in a cup nest is a matter for scientific debate.

CASE STUDY
American Yellow Warbler

The American Yellow Warbler builds a strong, well-attached nest, which is vulnerable to a specific form of parasitism. One of the most common of the 'wood warblers', the Yellow Warbler overwinters in Central America or northern South America and breeds right across North and Central America in mangroves, and in woods and thickets close to water.

Nest and nest site

In spring, the male chooses and defends the pair's territory. The female chooses the nest site about 1–2.4m (3–8ft) up, occasionally higher, and she alone builds the nest from plant fibres, dry grass, weed stems, down and wool. The base of the nest is attached firmly at the bottom of a fork of several uprights, especially in an alder or willow tree. The cup is built up with a succession of entangled stems and fibres. The bird twists and turns its body to shape the cup as the nest grows in size. She ties the top lip of the cup with loops of grass around the uprights, securing the cup firmly to the tree. The nest is lined with plant down, such as cotton, and maybe some fur and a few feathers.

Eggs and young

The female incubates the eggs. The male brings her food and later also brings food for the chicks. Typically, out of 4–5 eggs in a clutch, only 55 per cent of nests raise one or more young.

Predators and parasitism

Snakes and crows are potential dangers to the nests, while northern populations are the regular host of the parasitic Brown-headed Cowbird (*Molothrus ater*). Despite the mobbing behaviour of the warblers, as many as 40 per cent of nests suffer attempted or successful parasitism. If the warbler recognizes the cowbird egg, it will often cover it and its own eggs with new nest-lining material, and then lay a replacement clutch on top. Occasionally, if they are not covered, all the eggs hatch, and the mixed brood lives together and fledges successfully.

Classification

ORDER	Passeriformes
FAMILY	Parulidae
SPECIES	*Dendroica petechia*
RELATED SPECIES	Over 100 other wood warblers
NEST TYPE	Cup-shaped
SPECIES WITH SIMILAR NESTS	Many among songbirds
NEST SPECIALIZATION	New lining over parasite eggs

The nest is gradually built up with stems and fibres.

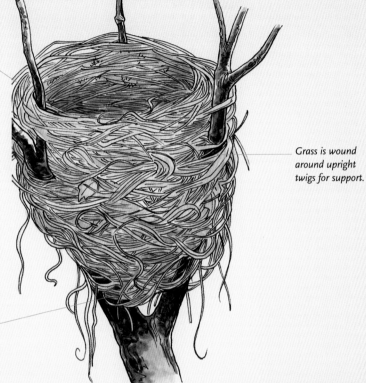

Grass is wound around upright twigs for support.

The nest base is secured in the tree fork.

SECURE ATTACHMENT
The very sturdy nest of the American Yellow Warbler is about 7.5cm (3in) across. The cup is firmly attached to the upright twigs by strands of grass wound around several stems, so the whole cup is quite rigid in the tree or shrub. Sometimes the nest is in a fork growing directly from the trunk. This means that the nest actually leans on one side against the trunk.

SOFT LINING
The thickness of the cup's lining gives an indication of its softness as it holds the eggs. Three feathers are curled around on the cup's lining, which is principally of cotton and coloured plant down. The clutch of four eggs shown here is typical of the widespread North American subspecies; the 'Mangrove Warbler' subspecies of the Caribbean lays only three eggs.

Chaffinch

The Chaffinch offers a good example of a well-camouflaged cup nest anchored with spider silk. This bird is widespread in Europe; many are resident, but northern and eastern birds migrate south and west. Living in woodland, farmland hedges, orchards and gardens, the male defends territory and helps feed the young; the female builds the nest. This is used once, but material may be reused in a new site.

Nest materials

The female builds the nest alone in about a week; the process can take as few as 3 days or as many as 18 days. She builds a neat, deep cup, tightly fitted into a shrub or tree fork, of moss, lichen, grasses and feathers, tethered with spider silk.

Nest structure

The nest cup has four layers. Strands of silk (attachment) join the cup to the branch. The outer layer (decoration) of lichen and spider silk looks green or greyish, providing camouflage. The cup (structure) is built with fine grasses and moss, in which grasses are prominent. The fourth element (lining) consists of a few feathers, fine grasses, wool, fur and plant down. The soft materials adhere well together so that many nests blend with lichened or mossy branches and are hard to find. The cup is strong and flexible enough to expand as the chicks grow.

Technique

Peter Marler described in detail exactly how the female builds. She uses two movements to attach the spider webs. First, a thread is attached to a branch; the leaning bird passes it around the branch, retrieves the end and so encircles the twig; she may make several loops in this way. Second, moss is stuck to the loops and more silk may be attached and secured elsewhere. Moss and grass are then inserted into the cup by stabbing movements of the bill. The whole is shaped from the inside as the bird turns around and around, making the cup a perfect fit for the incubating bird.

Classification

ORDER	Passeriformes
FAMILY	Fringillidae
SPECIES	*Fringilla coelebs*
RELATED SPECIES	Brambling, Blue Chaffinch and cardueline finches
NEST TYPE	Cup-shaped
SPECIES WITH SIMILAR NESTS	Many among songbirds
NEST SPECIALIZATION	Spider-web anchors; camouflage

LEFT
SPIDER-SILK ATTACHMENT
These Chaffinch chicks are begging for food in a nest in Gillenfeld, Vulkaneifel, Germany. The nest materials of moss, grass and lichen are clearly visible, as are the very fine strands of spider silk that attach the nest securely to the site.

CASE STUDY
American Robin

Mud cement plays an important part in the cup nest of the American Robin. Many of these well-known members of the thrush family winter in the United States, but others migrate in flocks as far south as Central America, and return to Canada by early March. They breed across North America in forests, farms and gardens, up to the tree line, in territories that may be as large as 0.13ha (⅓ acre).

Nest site

Males generally arrive first in the same general area as the previous year. They claim and defend a territory, marking their presence with rich, continuous singing. Females usually arrive about a week later. The pair choose the nest site together. Typically it is a tree fork (as shown here). Observers have recorded nearly 80 species of trees and shrubs as nest sites, and many strange ones, such as on top of another species' nest, on a building ledge and on a railway wagon.

Nest building

The nest is usually 3m (10ft) up in the bush or tree, but ground level and 19m (62ft) high have been recorded. The bulky, rather untidy nest is built entirely by the female, who makes on average 180 trips per day for 2–6 days. The first an observer might notice of the female is her interest in puddles of mud. The cup-shaped nest starts as a rough platform of grass and twigs. The mud is layered on it and worked in place with bill and feet. More grass is pressed in and more mud is added until the full-size cup is formed, which is thinly lined with fine grass. The nest measures up to 16.5cm (6½in) in diameter, with a cup depth of 6.5–10cm (2½–4in).

Eggs and young

The female usually lays four unmarked blue eggs, which she incubates for 11–14 days. Chicks are fed in the nest by both parents for about two weeks, and by the male for several days afterwards while the female starts a second clutch. The young stay in the territory for about three weeks before dispersing. A third brood is not unusual as late as August.

Classification

ORDER	Passeriformes
FAMILY	Turdidae
SPECIES	*Turdus migratorius*
RELATED SPECIES	Many in the same family on every continent
NEST TYPE	Deep, solid cup
SPECIES WITH SIMILAR NESTS	Songbirds of similar size
NEST SPECIALIZATION	Mud cement

CASE STUDY
Carrion Crow

The Carrion Crow of western Europe builds a new, sturdy nest each year, with a loose structure disguising its cup shape. Mainly resident, the Carrion Crow breeds in open woodland, scattered trees on moorland, farmland hedges, and city parks and gardens. Paired birds stay together all year, aggressively defending their breeding territory of around 9.7ha (24 acres) against other crows during incubation.

Nest site

Carrion crows generally build in the top or middle of the crown of a tree belonging to the region's most common species – for example, oaks in northern Germany, conifers in Scotland and Denmark, and poplars in Poland. The nest is almost always in the tallest tree in the territory, in a fork near the trunk, at an average height of 14–15m (46–49ft). Very rarely, nests have been found at ground level: in a Russian reed bed, and on a cliff ledge in a German seabird colony.

Nest building

It takes the male and female 1–2 weeks to build the nest. From a distance, or just from below, the nest looks more like a small platform nest of sticks, with the outer shape broken by many protruding ends that give a much looser shape than many in this category. Close inspection shows it is, indeed, a cup-shaped nest. The attachment is a foundation and wall of entangled twigs snapped off trees; a structure of bark strips, fine twigs, turf and moss is built up on top of this; and earth and mud are pushed into the bottom and some way up the sides to form a firm base. The lining consists of feathers, down, wool and even artificial materials. In sheep-rearing country, the thick, white lining of wool appears very striking against the dark twigs.

Nest size

The outer size varies with the type of sticks chosen, but the inner cup is often just 19–20cm (7½–8in) in diameter and 10–13cm (4–5in) deep.

Classification

ORDER	Passeriformes
FAMILY	Corvidae
SPECIES	*Corvus corone*
RELATED SPECIES	Crows of similar size in India, Africa and North America
NEST TYPE	Large cup-shaped
SPECIES WITH SIMILAR NESTS	Other crows
NEST SPECIALIZATION	Loose cup structure, wool lining

CHAPTER SIX

Domed Nests

Many species build a dome over the nest cup to form a style of avian architecture that incorporates some of the best features of the cup and of the cavity styles of nest. A domed roof provides protection from sun, rain and cold; reduces heat loss from a nest; and offers a secure and well-insulated refuge for eggs and young.

Domed nests are built by large and small birds, ground and tree nesters, and they vary greatly in terms of shape, size and materials. What might be called conventional domed, or ovoid, nests are built by wrens, some warblers, and Australian finches in low cover or on the ground. In contrast, the fortress nest of the Eurasian Magpie (*Pica pica*) is an irregularly shaped cup with a canopy of sticks and an inconspicuous opening to one side (see above, and page 72). The choice of style and alignment of entrance is an important architectural consideration. The Eastern Meadowlark (*Sturnella magna*) aligns its nest away from the prevailing winds to protect the chicks from the wind. Other domed nests have entrance tunnels that offer protection from predators such as snakes.

As with all nests, the protection and incubation of the eggs and young are the primary objectives, and dome architecture incorporates a variety of features to achieve these goals. As with cup nests, the chicks are nidicolous. They are fed in the nest by the parents until they are ready to fly, which among the passerines is usually from two to three weeks.

Choice of nest site is a key element in defence, and thorny sites can offer natural defences. The Cactus Wren (*Campylorhynchus brunneicapillus*) builds its nest in the prickly pear cactus to deter predators.

Some dome builders employ specialist defences. The Rainbow Pitta (*Pitta iris*) of northern Australia builds a domed nest on the rain-forest floor. As a protection against snakes, which depend on smell to find prey, the pittas collect smelly wallaby droppings and put them in the nest to disguise its presence. The Yellow-rumped Thornbill (*Acanthiza chrysorrhoa*) of Australia has evolved a closed nest of dry grass, feathers, plant down and spider web, with a roughly made cup-shaped 'false nest' built on top of the dome. The proper nest entrance is concealed with a hood. When a currawong (*Strepera* spp.), an aggressive nest raider, sees an apparently empty nest, it moves on.

Despite such protective measures, domed nests can suffer from squatters and predation. Although they seem to offer better protection than cup nests, it has not been proved that they offer better defence against predators.

RIGHT
WOOD WARBLER NEST
A Wood Warbler (Phylloscopus sibilatrix) *at its ground-level dome among wood and leaf litter.*

BLUEPRINTS
Domed Nest Structures

The basic blueprint for the domed nest is a cup nest that is extended and enclosed. However, domed nests vary greatly in size and appearance – from a small sphere to the Hamerkop's (*Scopus umbretta*) enormous construction. Birds that construct domed nests include the magpies, New World warblers, Old World leaf warblers, wrens, scrub birds, some swifts, and dippers. The Long-tailed Tit (*Aegithalos caudatus*) creates one of the most skilfully constructed dome nests: it is oval with a round entrance, and is attached using spider silk and the impressive Velcro technique (see pages 76–77).

FIG. I
STICK CUP AND CANOPY
The nest of the Eurasian Magpie (Pica pica) is unusual for a member of the crow family. The basic structure is a substantial cup – like a crow's (see page 69) – of sticks and mud, but with a lining of rootlets, grass, leaves and other soft material. Its main feature is its dome, which is made by extending the outer layer of sticks over the cup, and incorporating small twigs of the nest tree, until there is one entrance. The nest is often in a thorny tree; that and the dome offer a good defence against raiding crows. Both sexes help to build the nest, which takes about three weeks to complete.

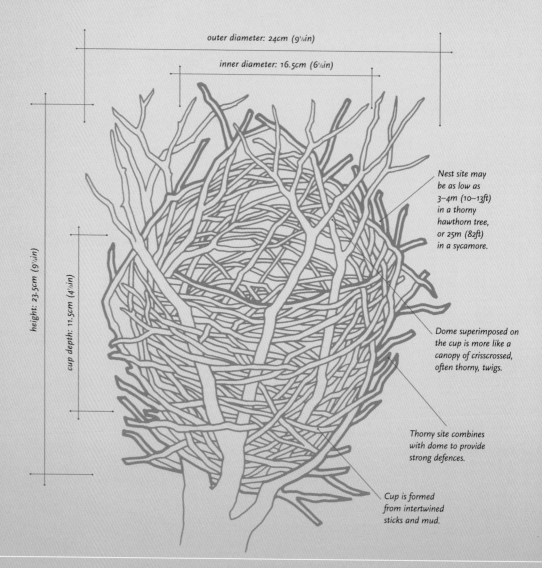

outer diameter: 24cm (9½in)

inner diameter: 16.5cm (6½in)

height: 23.5cm (9¼in)

cup depth: 11.5cm (4½in)

Nest site may be as low as 3–4m (10–13ft) in a thorny hawthorn tree, or 25m (82ft) in a sycamore.

Dome superimposed on the cup is more like a canopy of crisscrossed, often thorny, twigs.

Thorny site combines with dome to provide strong defences.

Cup is formed from intertwined sticks and mud.

FIG. I EURASIAN MAGPIE NEST

VARIETIES OF STRUCTURE

The dome may be an extension of sticks over the cup, as in the Eurasian Magpie's nest; a sphere of grass and feathers, as built by the Common Chiffchaff (*Phylloscopus collybita*); or an untidy dome of thorny twigs, as formed by the Verdin (*Auriparus flaviceps*). A typical leaf warbler's nest is built by the migrant Wood Warbler (*Phylloscopus sibilatrix*). As shown on page 71, it creates a slight depression in the ground and builds a completely domed nest. Some species use lavish amounts of lining materials.

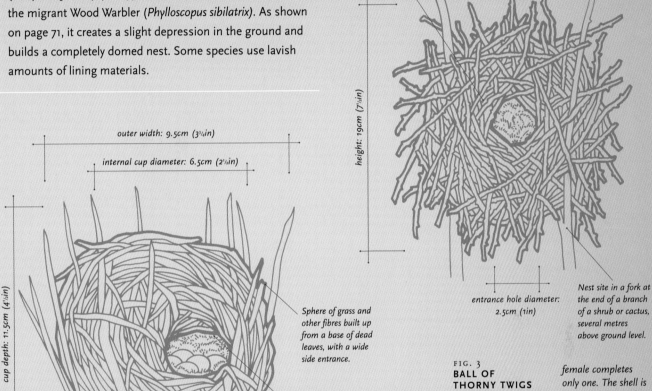

Ball-shaped nest with a round side entrance and thorn defence.

width: 14cm (5½in)

inside cavity diameter: 3.75–6.5cm (1½–2½in)

height: 19cm (7½in)

outer width: 9.5cm (3¾in)

internal cup diameter: 6.5cm (2½in)

cup depth: 11.5cm (4½in)

Sphere of grass and other fibres built up from a base of dead leaves, with a wide side entrance.

Luxury lining of feathers.

Site is often 50–90cm (½–3ft) above the ground.

entrance hole: 4.5cm (1¾in) wide 3.75cm (1½in) deep

entrance hole diameter: 2.5cm (1in)

Nest site in a fork at the end of a branch of a shrub or cactus, several metres above ground level.

FIG. 2
GRASS SPHERE, FEATHER LINING
Common Chiffchaffs breed in Europe and temperate Asia. They are closely related to the Wood Warbler and build a similar nest. The female alone constructs a well-concealed sphere in thick undergrowth, such as low brambles, just off the ground. The nest is often well lined with feathers; one nest in England was found to contain 670 chicken feathers and a few from local wild birds. Early nests in April are sometimes easy to see, but soon become well hidden as the herbage grows.

FIG. 3
BALL OF THORNY TWIGS
The Verdin is the only American member of a family that is widespread in southern Europe, Asia and Africa. It breeds in the southwestern United States and northern Mexico in arid regions of thorny scrub-like prickly mesquite or chaparral. Males start several nests but the female completes only one. The shell is an untidy mass of small thorny twigs, lined with grass, leaves, spider web and finally a thick layer of feathers and plant down. Nests for breeding are begun in late February. The young fledge at about three weeks old and continue to roost in the nest.

FIG. 2 COMMON CHIFFCHAFF NEST | FIG. 3 VERDIN NEST

Cactus Wren Nest

The Cactus Wren (*Campylorhynchus brunneicapillus*) is the largest wren in North America, residing in the southwestern United States and much of Mexico in arid regions of low scrub and cacti. Both the male and female build the nest almost exclusively in Prickly Pear cactus (*Opuntia* spp.) and Cholla (*O. prolifera*). The construction is a large, domed, oval, woven from twigs, dry grass, and rootlets. Viewed from the side, the nest looks elongated. On the outside, it looks untidy and is often very conspicuous. Young birds make a nest as soon as they are independent. A pair constructs nests all year round, one of which is chosen for breeding. Outside the breeding season, the nests are used as roosts. Roosting nests like these are very rare among birds. Generally, 3–4 eggs are laid, incubated by the female. The chicks are fed by both parents and fledge after around 21 days.

Ventilation

Temperature is all-important when incubating eggs. The entrance to the Cactus Wren nest is often orientated to take advantage of convective ventilation provided by prevailing winds. The nesting cavity is about 6in (15cm) in diameter, and is accessed through a tube of twigs 6in (15cm) long. The nest has thick walls, and although it looks coarse, the egg chamber is softly lined with fine grass, feathers, and plant down. So the eggs are both ventilated and incubated.

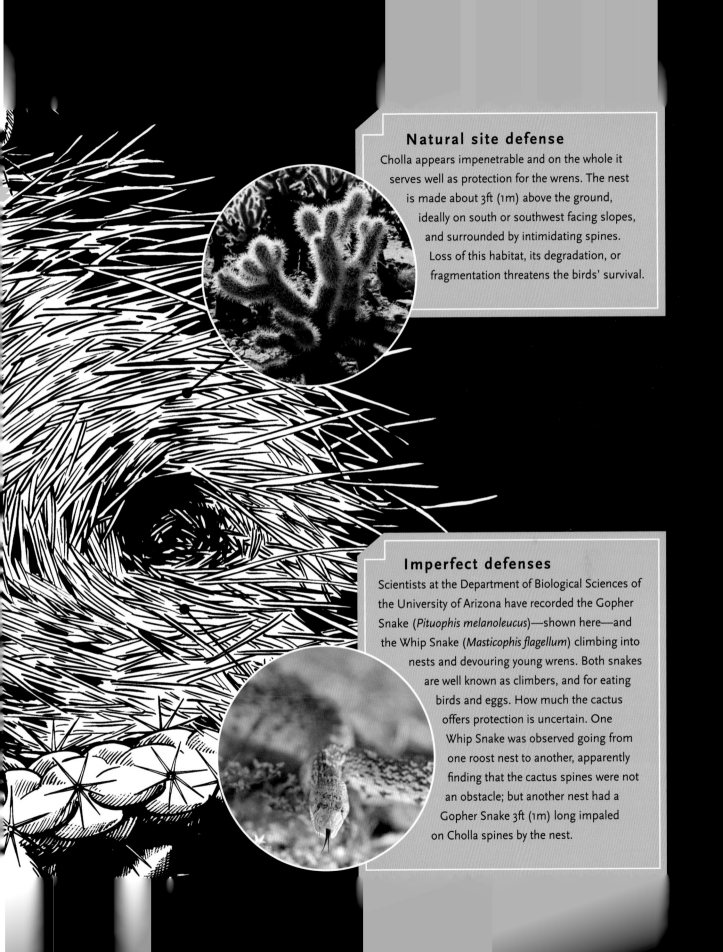

Natural site defense

Cholla appears impenetrable and on the whole it
serves well as protection for the wrens. The nest
is made about 3ft (1m) above the ground,
ideally on south or southwest facing slopes,
and surrounded by intimidating spines.
Loss of this habitat, its degradation, or
fragmentation threatens the birds' survival.

Imperfect defenses

Scientists at the Department of Biological Sciences of
the University of Arizona have recorded the Gopher
Snake (*Pituophis melanoleucus*)—shown here—and
the Whip Snake (*Masticophis flagellum*) climbing into
nests and devouring young wrens. Both snakes
are well known as climbers, and for eating
birds and eggs. How much the cactus
offers protection is uncertain. One
Whip Snake was observed going from
one roost nest to another, apparently
finding that the cactus spines were not
an obstacle; but another nest had a
Gopher Snake 3ft (1m) long impaled
on Cholla spines by the nest.

The Velcro Technique

Although it is called a 'tit', the Long-tailed Tit (*Aegithalos caudatus*) is not actually a member of the tit or chickadee family, but is related to the American Bushtit (*Psaltriparus minimus*) of North America and Mexico. The Long-tailed Tit's nest is one of the most beautifully formed and skilfully constructed nests in the world. Mike Hansell, of Glasgow University, Scotland, has established that to make a typical nest the birds collect around 3,000 lichen flakes, 600 or more silk spider egg cocoons, 200–300 sprigs of moss and about 1,500 feathers – although over 2,000 were counted on one nest. The pair working together take 9–14 days to build the shell, and another 1–2 weeks to collect and line the nest with feathers. The nest is a 13-cm (5-in) oval (even pear-shaped at times) with a neat, round entrance well above the halfway line towards the top. The nest walls are elastic and can expand as the 7–12 chicks grow. The moss and silk work like Velcro. The moss (the hooks) is attached to the tree fork with silk strands (the hoops) that can be unfastened and refastened to keep the nest tightly attached as it expands.

1. *The bird starts the nest by firmly placing moss in the fork of a bush or creeper 1–4.5m (3–15ft) high, and 'tying it in' with silk.*

2. *As the moss cup grows, the bird uses its bill to embed flakes of lichen into the moss on the outside of the shell.*

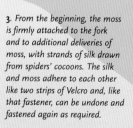

3. *From the beginning, the moss is firmly attached to the fork and to additional deliveries of moss, with strands of silk drawn from spiders' cocoons. The silk and moss adhere to each other like two strips of Velcro and, like that fastener, can be undone and fastened again as required.*

LONG-TAILED TIT DOME
An adult Long-tailed Tit at its nest, here partly concealed among bamboo in Central Europe.

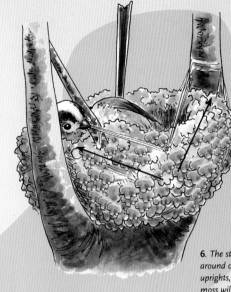

6. *The strand is turned around one of the fork's uprights, to which more moss will be attached. This process continues until the nest's shell is complete.*

5. *A strand of silk is drawn from a cocoon in the moss and lichen.*

4. *The builder shapes the growing 'ball' by sitting in it and turning around and around.*

SPECIALITY MATERIALS COLLECTOR
This drawing shows loops of spider cocoon silk around small leaflets of moss. The silk and the moss build up a wall, held together by the 'Velcro principle'. The moss is mostly from two small-leaved branching species. (Based on a microphotograph by Margaret Mullin.)

CASE STUDY
Eastern Meadowlark

The meadowlark builds a well-concealed dome aligned to provide protection from the wind. Eastern Meadowlarks (which are not larks but icterids, related to grackles and North American orioles) are found from southeast Canada south to the Gulf Coast, and westward to the central plains. They are widespread in open grassland, prairie, pastures and hay fields. They feed on insects as they forage on the ground.

Nest and nest building

The male is polygamous, and has two or three mates. The female builds the nest while the male defends his territory, which ranges in size from 1.2–6ha (3–15 acres). She builds the nest on the ground in a shallow depression, so well concealed in thick groundcover that a person can easily pass within a metre without noticing the nest. The nest is made of dry grass and plant stems, lined with finer grass and some hair. There is a dome of grass over the cup; many of the strands that the female brings are entwined in the long grass around the nest. There is a large, elongated side entrance. The closely related Western Meadowlark (*Sturnella neglecta*) builds a very similar nest, and the visits of both species to their nests often result in a small tunnel being formed in the grass.

Young

The male will help feed the young, and may do so on his own when the young fledge and the female makes a new nest.

Defence, parasitism and predators

The dome offers only partial defence. Brown-headed Cowbirds (*Molothrus ater*), which are in the same family, sometimes parasitize Meadowlark nests. Nests also suffer from predation, especially if they are not fully covered with a grassy canopy. It was discovered by Roseberry and Klimstra (1970) that 60 per cent of nests without a canopy were lost to predation, compared to 51 per cent of nests that had been completely covered to conceal the eggs.

Classification

ORDER	Passeriformes
FAMILY	Icteridae
SPECIES	*Sturnella magna*
RELATED SPECIES	New World blackbirds, orioles, caciques, grackles and cowbirds
NEST TYPE	Domed ground nest
SPECIES WITH SIMILAR NESTS	Some Old World warblers
NEST SPECIALIZATION	Wind protection

Direction of wind – the nest is positioned to avoid wind blowing into the entrance.

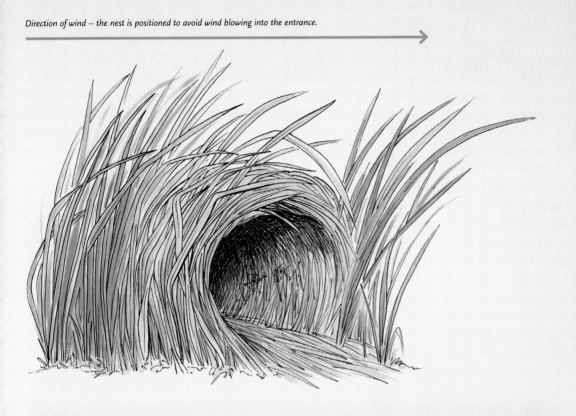

WIND PROTECTION
Scientists at two universities in Kansas studied 272 Eastern Meadowlarks' nests and discovered that they were not built randomly with regard to the orientation of the entrance. Generally, the entrances faced northeast, away from the prevailing southerly winds. They thought that this might give the birds a structural building advantage, protecting the chicks from the wind, as well as keeping them in the shade; and as the prevailing wind shifted southeast later nests tended to face more northward.

CAMOUFLAGE AND CHICKS
Set deep in the long grass from which its dome is constructed, the Eastern Meadowlark nest is always hard to find. But as this photograph shows, the mother will have no problem identifying the needs of her young. The large entrance and the dome of grass frame the chicks being fed by the adult at Scarborough, Ontario, Canada, near the northern limit of the species' distribution.

CASE STUDY
Eurasian Wren

The Eurasian Wren builds secure, snug and well-concealed domed nests using the vegetation that surrounds the site. Wrens are widely distributed in a variety of habitats from woods to farmland, on offshore islands, on moorland and in city suburbs. There are 80 or so species of wren in the world. The American Winter Wren (*Troglodytes hiemalis*) is a close relative of this Old World species.

Males and multiple nests

The male starts the breeding season by building an average of 6 nests, or sometimes as many as 12, which are commonly known as 'cock nests'. Often one or more of these is left unfinished. He builds in a hollow, crevice or hole, in a tree, steep bank, nest box or a garden shed. If constructed in a bush, creeper or low vegetation, the nest is domed with a side entrance; those built in a crevice, bank or wall may consist of a cup alone. Whatever the site, it will provide a snug position, concealment, good secure points to which the nest material can be attached, and a clear flyway from the nest hole. The nest also serves as a good place for the male's advertisement display and song to attract a female.

Nest building

The nest is made of grass, moss, dead leaves and other vegetation. In his book *The Wren* (1955), Edward A. Armstrong described how during the first few visits in late March the bird lays a firm base in a fork or on a ledge. A thin roof may be added while the shell is still flimsy, but eventually the inside is 'upholstered', as Armstrong put it, with moss. The threshold of the entrance hole is often strengthened by weaving in dry grass. A nest can appear to be finished in a day, but it takes 4–5 days to complete it.

Female nest lining

Once a female has been attracted to a nest, she lines it with a 'mattress' of feathers, down, hair or fur (for example, from deer or foxes). This takes 2–7 days; she sometimes even adds material while the 5–8 eggs are being laid.

Classification

ORDER	Passeriformes
FAMILY	Troglodytidae
SPECIES	*Troglodytes troglodytes*
RELATED SPECIES	About 60 other wrens, dippers
NEST TYPE	Domed
SPECIES WITH SIMILAR NESTS	Dippers, other wrens, some warblers
NEST SPECIALIZATION	Multiple nests built by male

LEFT
**EURASIAN WREN
DOME NEST**
*This typical Eurasian
Wren nest photo
captures the neatly
concealed domed nest
and the vociferous
demands of the
young. Moss, dead
leaves and grass
stems completely
fill the shape of the
cavity in the tree
trunk. The dome of
moss curves behind
the leaf and over the
heads of the well-
grown young. The
only thing we cannot
see is the lining of
feathers, which is sure
to be there. The
number of feathers
depends on their
availability; nearly
500 have been
counted, but other
nests are mostly lined
with hair or fur.*

CASE STUDY
Hamerkop

The Hamerkop, or Hammerhead, builds the largest domed nest of any bird, a structure that forms a conspicuous part of the African landscape. Found throughout Africa south of the Sahara and the southwest corner of the Arabian Peninsula, this bird is in a family of its own. Hamerkops are never far from water — where they get their food — and trees where they roost and nest. The birds are believed to pair for life.

Nest and nest site

Hamerkop pairs make their nest in a fork of a tree with a bowl of twigs, reeds and grass cemented with mud. The first sticks of the dome are placed diagonally across the corners, then the birds work inward until the dome is complete — but not finished off. The sides are extended and roofed over with sticks and branches 60–80cm (2–2⅔ft) long and 2.5–3.75cm (1–1½in) thick.

Nest features

A thatch of plant debris features a range of other unusual materials — snake skins, refuse, dung and carrion are added. The mud helps to insulate the nest and make it waterproof and predator-proof. The single entrance is well hidden, low on one side and leads into a 13–18-cm (5–7-in) diameter, mud-lined tunnel, up to 60cm (2ft) long. There are divisions or ledges inside but there are no strictly designated 'rooms'. A new nest measures about 90cm (3ft) wide by 1.5m (5ft) high. It may take 6–8 weeks to complete and contain around 10,000 sticks. It is used in the following years, and more sticks are added to the outer shape until it may be 2m (6½ft) wide and deep, and solid enough to withstand a man standing on it.

Squatters and companion nests

Large owls sometimes drive away the Hamerkops and take over the nest, but the builders have been known to reoccupy it again after the 'squatters' have left. Weaverbird and pigeon nests have been discovered built onto the nest, while snakes, small mammals and other birds live in abandoned Hamerkop nests.

Classification

ORDER	Ciconiiformes
FAMILY	Scopidae
SPECIES	*Scopus umbretta*
RELATED SPECIES	Herons, storks, ibises, spoonbills
NEST TYPE	Domed stick structure
SPECIES WITH SIMILAR NESTS	None similar
NEST SPECIALIZATION	Large size

CASE STUDY
Firewood-Gatherer

The appropriately named Firewood-gatherers are only 19.5cm (7¾in) long, yet they construct a huge domed nest. Their building technique involves a painstaking process of gathering sticks one by one. Resident in central and southeast Brazil southward to Uruguay, Paraguay and northern Argentina, the Firewood-gatherers inhabit acacia savanna, grassland and agricultural land with scattered trees.

Classification

ORDER	Passeriformes
FAMILY	Furnariidae
SPECIES	*Anumbius annumbi*
RELATED SPECIES	Thornbirds, and other members of the ovenbird family
NEST TYPE	Stick dome
SPECIES WITH SIMILAR NESTS	Spinetails or castle-builders in the same family
NEST SPECIALIZATION	Large structure, spiral entrance

Nest site and materials

The Firewood-gatherer builds its nest in an isolated tree but may use a telephone pole. It gathers sticks one by one, holding the stick in the middle to balance it, and places it on a crook of branches. Many sticks, however, fall to the ground, making a pile below the tree. The bird is unable to pick up a fallen stick since it cannot rise vertically carrying the load. Therefore, when another stick is needed, the bird flies off for a new one. Eventually a hollow 'tower' about 60cm (2ft) tall and 25–30cm (10–12in) wide is formed, with an entrance at the top. It sits at an oblique angle on the branches. The great Victorian naturalist W. H. Hudson, who lived in Argentina for many years, described the nest building as 'a most laborious operation, as the sticks are large and the birds' flight is feeble'. This is very understandable for a bird whose mean weight is only 41.5g/1½oz (Dunning, 1994).

Nest structure

The finished nest is drum-shaped with an entrance at the top from which a crooked or spiral passage leads down to the off-centre breeding chamber. This is lined with wool and soft grass. The crooked passage helps to prevent rain from entering the breeding chamber. The Firewood-gatherer lays five white eggs in the chamber.

Family roosting

The whole family of Firewood-gatherers stays together for months after the young have fledged, using the nest as a roosting place.

CHAPTER SEVEN

Mud Nests

Not content with a house of straw, the potters of the avian world have evolved to build solid and durable architecture in which mud is the dominant material. 'According to the theory that men acquired their first notions of architecture from birds, we are told that Doxius, the inventor of clay houses, took the hint from swallows'. This was recorded in Pliny the Elder's famous *Natural History* of 77–79 CE.

Mud is crucial in nest building for only about 5 per cent of species, and only a small proportion of those build their nests completely or predominantly from mud. These few species utilize the technique of moulding to obtain a variety of shapes, and have developed a number of other skills specific to this admirably durable and adhesive material.

A strong motivation for using mud as construction material is its structural convenience – it can be moulded into the required shape and space. As with domed nests, the original structure seems to be the cup, as shown above in the nest of a Barn Swallow (*Hirundo rustica*) of the Northern Hemisphere. Birds have adapted to build up the cup into an enclosed and well-protected shelter with an entrance.

Adhesion is a key requirement. Mud nests are built on different substrates including vertical surfaces such as walls or even the ceilings of caves, and many are heavy and have no support beneath them. Wet mud can form a strong glue, sticking even to vertical surfaces, and birds add their own saliva to the list of construction materials, which seems to have the effect of increasing the mud's adhesive qualities.

As an example, the White-necked Rockfowl (*Picathartes gymnocephalus*) is able to attach its nest to a vertical rock face, an especially impressive feat since the nest weighs 2kg (4½lb), the heaviest of any rock-attached mud nest.

Like a potter, the nest builder faces two problems: air trapped between successive loads and mud drying too soon, both of which can cause weaknesses in the structure. To address this, mud builders vibrate their heads when applying a new beakful of mud, which distributes moisture and moulds the new load onto and into the drier surface. In order to avoid the collapse of the heavy nest from the wall, birds pause between bouts of building to allow for the mud to dry and harden. This work can be seen in a nest's banded appearance. Once dry, mud is extremely durable, and additional plant materials are added for strength.

A disadvantage is that water is needed to make mud, which restricts the time and location for building. It can take the White-winged Chough (*Corcorax melanorhamphos*) several months to construct its nest if there is a lack of rain. Nests are durable enough to be reused, but this can also encourage infestation by bloodsucking nest ectoparasites.

RIGHT
WESTERN ROCK NUTHATCH
The Western Rock Nuthatch (Sitta neumayer) breeds in rocky, mountainous regions from Serbia to Iran. This nest is in a rock crevice and features an inner cup of moss, hair and feathers, sealed in with a cone-shaped mud front.

BLUEPRINTS

Mud Nest Structures

Nests moulded primarily from mud follow the basic blueprint of a cup, which is then built up with pellets of mud to form a variety of structures from half-cup to enclosed retort. The choice of substrate ranges from a horizontal branch to a vertical cliff wall to the ceiling of a cave. In the latter cases, with no support from beneath, strong adhesion is essential. The sequence of building is typically initial adhesion; piling up and moulding of mud pellets; plastering of rim; and lining. Mud builders are relatively scarce and include swallows, martins, flamingos and Magpie-larks (*Grallina cyanoleuca*).

FIG. I
MUD PELLET HALF-CUP

The Common House Martin (Delichon urbicum) is well known in Europe for attaching its rounded half-cup of mud to the walls of houses, just beneath the eaves. The nest chamber is enclosed except for a small entrance at the top, and is built by both parents. It is lined with feathers and plant material. The nest is built up from the bottom with pellets of wet mud collected from up to 45m (50yd) away, and secured in place with a rapid shivering movement of the head and bill. About 1,500 pellets are transported, one by one, in the bill. It breeds in loose groups of several birds, and also in colonies of hundreds with nests touching each other.

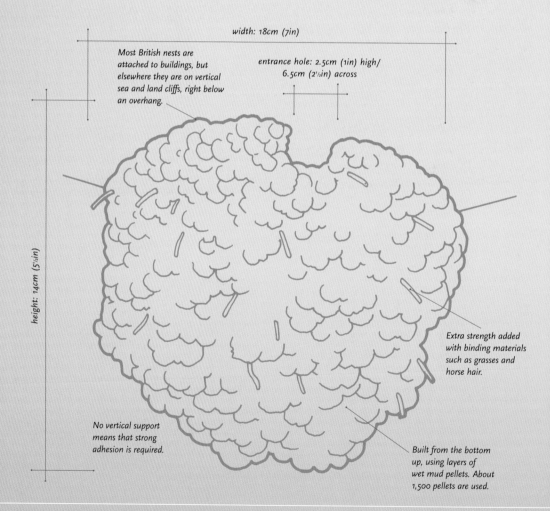

width: 18cm (7in)

Most British nests are attached to buildings, but elsewhere they are on vertical sea and land cliffs, right below an overhang.

entrance hole: 2.5cm (1in) high/ 6.5cm (2½in) across

height: 14cm (5½in)

Extra strength added with binding materials such as grasses and horse hair.

No vertical support means that strong adhesion is required.

Built from the bottom up, using layers of wet mud pellets. About 1,500 pellets are used.

FIG. I COMMON HOUSE MARTIN NEST

VARIETIES OF STRUCTURE

A cup-shaped nest is built by the Barn Swallow (*Hirundo rustica*) of the Northern Hemisphere; retort-shaped nests by the Fairy Martin (*Petrochelidon ariel*) of Australia; while the Western Rock Nuthatch (*Sitta neumayer*) builds a flask-shaped structure with an entrance tunnel up to 10cm (4in) long. This chamber is lined with soft materials. It is mostly built by the male, taking 10–18 days, and might be used for several years. The completed nest may weigh over 32kg (70lb), while the bird weighs only about 35g (1¼oz).

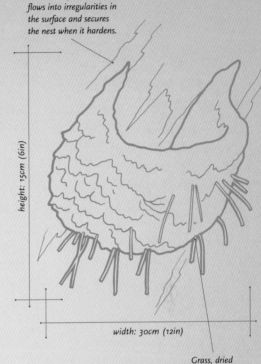

A cement of wet mud flows into irregularities in the surface and secures the nest when it hardens.

height: 15cm (6in)

width: 30cm (12in)

Grass, dried leaves and twigs are embedded in the nest.

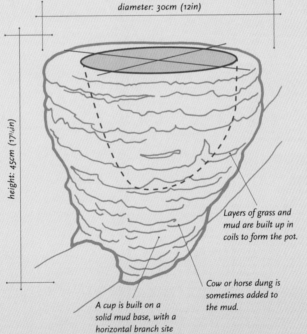

diameter: 30cm (12in)

height: 45cm (17¾in)

FIG. 2
MUD AND GRASS COIL POT
The White-winged Chough (Corcorax melanorhamphos) is a large bird that nests in woodland and scrub in southern and eastern Australia. Located high in a tree, the nest is built of mud and plant fibres in four stages: 1.) a solid base of mud is formed over a horizontal branch; 2.) a flat circular cup is added, firmed with

vibrating movements of bill and head; 3.) walls 2.5cm (1in) thick of mud and grass are raised in eight or more circular

strips of grass and mud mix, like a potter making a coil pot; 4.) the cup is lined with fine grasses and a few feathers.

It may take several months to build if there is not enough rain to soften the mud, but usually only several days.

Layers of grass and mud are built up in coils to form the pot.

Cow or horse dung is sometimes added to the mud.

A cup is built on a solid mud base, with a horizontal branch site providing vertical support.

FIG. 3
ADHESIVE MUD BOWL
The White-necked Rockfowl (Picathartes gymnocephalus) is a rare, crow-size West African bird living colonially, usually 2–5 pairs together, but sometimes as many as 40. Its substantial mud nest is constructed on the roof of a cave or a vertical rock face in

mountainous country, so adhesion of the nest to the surface is a key requirement. This is achieved with a glue of mud and saliva, likely with the hook and eye (Howard, 1997) process: the wet mud flows into crevices in the surface and as it dries and hardens it secures the nest in position.

FIG. 2 WHITE-WINGED CHOUGH NEST | FIG. 3 WHITE-NECKED ROCKFOWL NEST

MATERIALS AND FEATURES
Magpie-Lark Nest

The Magpie-lark (*Grallina cyanoleuca*) is one of Australia's
most widespread and best-known birds. It is neither a
magpie nor a lark but is related to monarch flycatchers.
Found in most habitats except forests, the Magpie-lark is
a regular in suburban areas. It commonly builds its nest
13–39ft (4–12m) above the ground on a horizontal surface on
a level tree branch, a telephone pole, a building, or a desert
windmill. The nest is a good example of a mud bowl that is
bound together and strengthened with plenty of dry grass
and other materials, and it is lined with fine, dry grass and a
few feathers. Breeding time varies with the rainfall: building
cannot start without a supply of mud. Robust and weighty,
the nest is rarely damaged while in use. A nest often
lasts into another year, and can be used by other
birds, such as honeyeaters. The Magpie-lark is a
favorite host of the Asian Koel (*Eudynamys
scolopaceus*), a migratory cuckoo.

Mud dwelling

Mud has been used by
humans for building since
ancient times, and can
last for thousands of years
(this African roundhouse is
built by hand using layers of
mud). Avian architecture with
thick walls of mud demonstrates
the same advantages as human
dwellings. Strong, durable, adhesive,
readily available, mud also helps to stabilize the
all-important temperature of the nest. Thick mud
walls absorb the heat of the sun during the day,
and slowly release it at night.

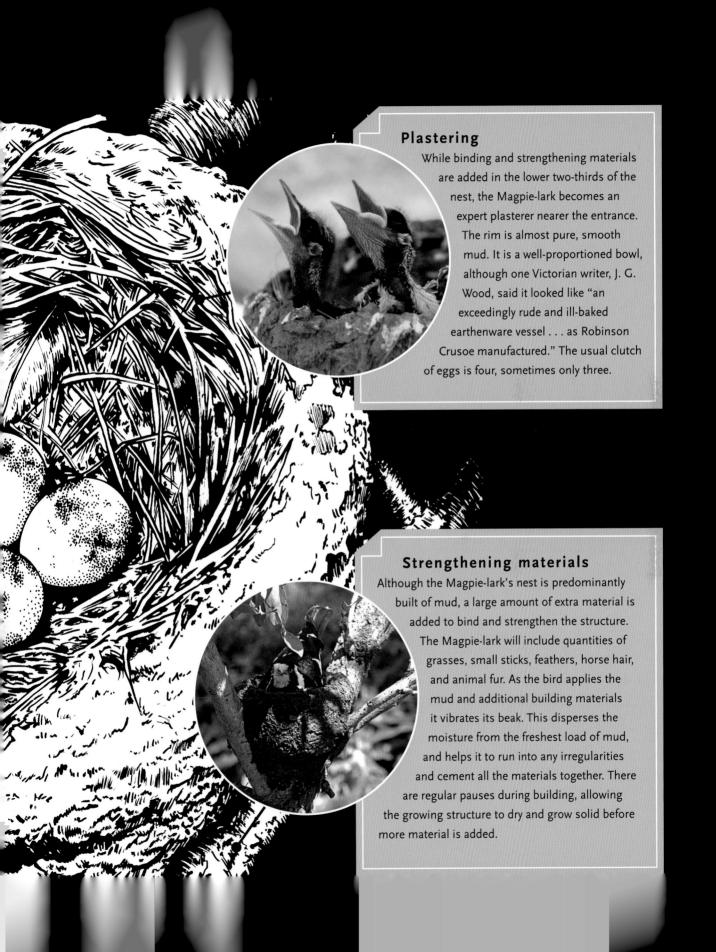

Plastering

While binding and strengthening materials are added in the lower two-thirds of the nest, the Magpie-lark becomes an expert plasterer nearer the entrance. The rim is almost pure, smooth mud. It is a well-proportioned bowl, although one Victorian writer, J. G. Wood, said it looked like "an exceedingly rude and ill-baked earthenware vessel . . . as Robinson Crusoe manufactured." The usual clutch of eggs is four, sometimes only three.

Strengthening materials

Although the Magpie-lark's nest is predominantly built of mud, a large amount of extra material is added to bind and strengthen the structure. The Magpie-lark will include quantities of grasses, small sticks, feathers, horse hair, and animal fur. As the bird applies the mud and additional building materials it vibrates its beak. This disperses the moisture from the freshest load of mud, and helps it to run into any irregularities and cement all the materials together. There are regular pauses during building, allowing the growing structure to dry and grow solid before more material is added.

Oven-Building Technique

OVEN CONSTRUCTION

This sequence shows the stages in building the Rufous Ovenbird's eponymous structure. It usually takes 2–3 months but can be done in about two weeks. The nest is well insulated from the heat of the day, but, unfortunately, the nest chamber becomes a good breeding ground for bloodsucking bugs, which sometimes infest the chicks. On the other hand, the indirect entry makes it very difficult for predators to get in, and the smooth, concrete-hard mud repels attacks.

According to ancient South American myth, it was the Rufous Ovenbird (*Furnarius rufus*) that taught people to build houses. Found from Brazil to Argentina in open country, this elegant brown bird builds spectacular nests. These structures are known in Spanish as *hornos* – 'ovens' – because that is exactly what they resemble. The birds pair for life and build the nest together using mud pellets with added materials. The nest is long lasting but not usually used again by the original pair. It may seem extraordinary that the builders do not make more use of their creation. The unused nest is often taken over by the Saffron Finch (*Sicalis flaveola*), a tanager, which is a cavity nester.

1. *The Rufous Ovenbird selects a level site such as a horizontal branch, tree stump, fence post, or ledge on a building. Both birds in the pair fetch and carry mud pellets one at a time in the bill.*

2. *The birds apply the initial adhesive base by building up layers of mud pellets. These are mixed with other materials such as cow dung, hair and straw to strengthen the structure.*

3. *The mud pellets are built up to form the thick walls of the cup.*

LEFT
**RUFOUS
OVENBIRDS
ON NEST**
*A life-long pair of Rufous
Ovenbirds sing at their
nest in the Pantanel
region of Brazil.*

6. *The finished 'oven'
is twice the size of a
football and weighs
about 4.5kg (10lb) – the
bird itself weighs only
57g (2oz). Durrell (1956)
said of one nest in
Argentina: 'The inside
of the little room and
the passageway was
smooth and almost
polished. The more
I examined the nest,
the more astonished
I became that a bird,
using only the beak
as a tool, could have
achieved such a
building triumph'.*

5. *A wall is built inward
to about three-quarters
of the height of the shell,
forming an entrance
lobby and side entrance
that help to keep the
nest chamber ventilated.*

4. *Over 2,000 pellets are
used to construct walls thick
enough to absorb the heat
of the sun. The ovenbirds
smooth the insides of
the moulded walls with
sculpting movements of
their beaks.*

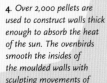

CASE STUDY
American Cliff Swallow

The American Cliff Swallow summers over much of North America to breed. It winters as far south as Argentina. Colonies usually number a few hundred nests, but 2,000–3,000 nests have been known. The 'wild' site for a nest is a cliff, but swallows also build on dams and bridges. They build durable gourd-shaped nests that can be reused in following years, although there is a danger of parasitic infestation.

Nest and materials

Pairs of cliff swallows build the nest together. Mud gathering is an animated and social activity: little groups fly to sources of mud, land for a few seconds, then fly off after forming a pellet in the bill. The birds work like this for a few hours each day for a week, or more if the weather is wet. As the nest grows, it is aggressively guarded by one of the pair because mud stealing by near neighbours is a serious threat.

Nest building

First, one bird – usually the male – cements a narrow crescent of mud to the rock or wall. This is then built up into a ledge on which the birds can stand. They make a cup 15–18cm (6–7in) across with a back wall; side walls are raised; a roof is added and finally the nest's tubular entrance, 15–20cm (6–8in) long, is formed. Some nests are left with a wide entrance. The nest is lined with dry grass and feathers.

Progressive parasitism

The same colony is often used in successive years, with damaged nests being repaired, but sometimes sites are changed to avoid parasitic Swallow Bugs (*Oeciacus vicarius*). The bigger the colony the worse an infestation can be. A 10-day-old nestling in a big colony has been found to lose one-sixth of its weight to bloodsuckers, and instead of the expected 70 per cent of eggs laid producing fledglings, only half the normal number of young survive. This infestation can result in a progressive buildup of parasites over several years until the birds are forced to abandon the colony and found a new one at a different site.

Classification

ORDER	Passeriformes
FAMILY	Hirundinidae
SPECIES	*Petrochelidon pyrrhonota*
RELATED SPECIES	Over 70 other swallows and martins
NEST TYPE	Enclosed mud
SPECIES WITH SIMILAR NESTS	Some other swallows and martins
NEST SPECIALIZATION	Colonies of mud gourds

LEFT
**NEAR
NEIGHBOURS**
*This cliff colony
clearly shows how
closely packed the
nests are – note how
the entrance tunnels
point downward and,
if possible, away
from a neighbour's
nest. This colony is
in Baca County in
Comanche National
Grassland, Colorado.*

CHAPTER EIGHT

Hanging, Woven & Stitched Nests

Among the weavers and tailors, avian architecture achieves its most sublime expression of nest-building skill. Most of the nests described in this chapter are 'pensile' or hanging, supported from either the top or the side. Without support from below, both attachment and construction rely on elaborate binding, weaving and knotting to create a secure nest. This produces some of the most extraordinary constructions in the natural world.

The true weavers are the African weaverbirds and New World oropendolas, caciques and orioles. These architects use identifiable knots and stitches to suspend and construct nests of grass materials that resemble woven baskets. As in human weaving, knots and weaves are repeated to create intricate structures. Woven nests are strong yet flexible, light yet highly resilient. To produce them, weaverbirds demonstrate a combination of gymnastics and engineering.

Weaving techniques to create a pensile nest vary in their complexity. The Village Weaver (*Ploceus cucullatus*), shown above, follows a complicated series of moves to weave strips of grass. The typical nest of the Baltimore Oriole (*Icterus galbula*) might contain 10,000 stitches. A great advantage of weaving is that it allows a bird to build a nest from one type of material. However, the bird must be skilled enough to ensure that the hanging structure does not fall apart. Although weaving birds are born with the instinct to create nests in this way, it is apparent that, at least in some species, their skill improves with practice.

Also featured in this chapter are the 'tailors' who use the naturally adhesive silk from spider webs to stitch together pensile nests. The Common Tailorbird (*Orthotomus sutorius*) makes a cone nest between the sides of a curled leaf, while the Goldcrest (*Regulus regulus*) sticks moss and lichen together with spider webs. The Sooty-capped Hermit (*Phaethornis augusti*) makes a cable from spider silk to attach her nest to its support. Several of Australia's honeyeaters (Meliphagidae) build a neat hammock of shredded bark and other dry materials, again tightly stitched together with spider silk.

Placed at the extreme edge of a tree or bush, hanging, woven and stitched nests make it extremely difficult for arboreal predators to reach them. Woven tunnel entrances of various forms add an extra layer of protection for the nest, eggs and young.

RIGHT
YELLOW-RUMPED CACIQUE'S WOVEN NEST
A Yellow-rumped Cacique (Cacicus cela) at its nest in the Pantanal region of Brazil, South America, weaving the bottom of the basket.

BLUEPRINTS

Hanging & Woven Nest Structures

Old World weavers and New World orioles, caciques and oropendulas create true woven nests. All pull fresh, flexible fibres from large grasses, banana leaves or similar. The birds knot, twine and weave the grasses around their chosen nest supports in order to create wads as a base from which to build. The oval entrance of a simple woven nest leads directly to a central chamber. This blueprint is adapted to globular, kidney, retort and purse shapes with a variety of tubed entrances.

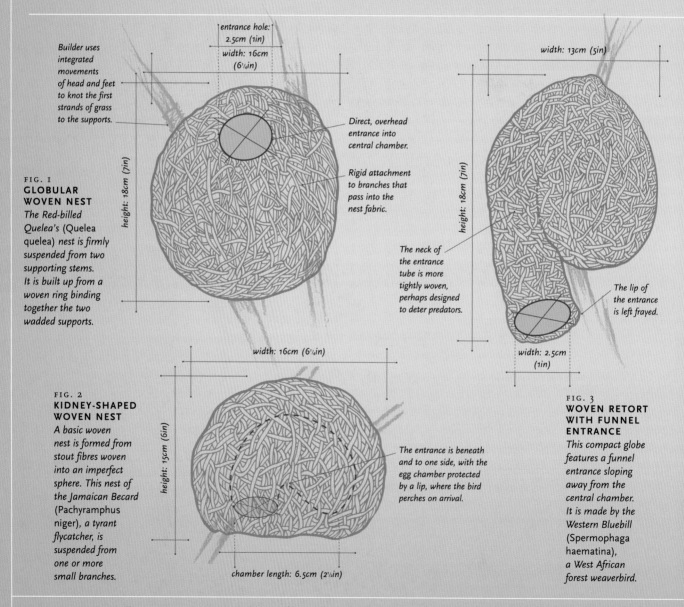

Builder uses integrated movements of head and feet to knot the first strands of grass to the supports.

entrance hole: 2.5cm (1in)

width: 16cm (6¼in)

width: 13cm (5in)

Direct, overhead entrance into central chamber.

Rigid attachment to branches that pass into the nest fabric.

height: 18cm (7in)

height: 18cm (7in)

FIG. 1
GLOBULAR WOVEN NEST
The Red-billed Quelea's (Quelea quelea) nest is firmly suspended from two supporting stems. It is built up from a woven ring binding together the two wadded supports.

The neck of the entrance tube is more tightly woven, perhaps designed to deter predators.

The lip of the entrance is left frayed.

width: 2.5cm (1in)

width: 16cm (6¼in)

height: 15cm (6in)

FIG. 2
KIDNEY-SHAPED WOVEN NEST
A basic woven nest is formed from stout fibres woven into an imperfect sphere. This nest of the Jamaican Becard (Pachyramphus niger), a tyrant flycatcher, is suspended from one or more small branches.

The entrance is beneath and to one side, with the egg chamber protected by a lip, where the bird perches on arrival.

chamber length: 6.5cm (2½in)

FIG. 3
WOVEN RETORT WITH FUNNEL ENTRANCE
This compact globe features a funnel entrance sloping away from the central chamber. It is made by the Western Bluebill (Spermophaga haematina), a West African forest weaverbird.

FIG. 1 RED-BILLED QUELEA NEST │ FIG. 2 JAMAICAN BECARD NEST │ FIG. 3 WESTERN BLUEBILL NEST

VARIETIES OF STRUCTURE

More complex nests have tubular entrances and are retort shaped. A cup-shaped lining made of softer material sits within the basket of strong, hard fibres; some nests feature a more densely woven 'roofing' layer above the central chamber to function as waterproofing. The Red-headed Weaver (*Anaplectes rubriceps*) is the only weaver to use twigs, which are secured to the nest with the help of tags of bark left from where the twig was detached.

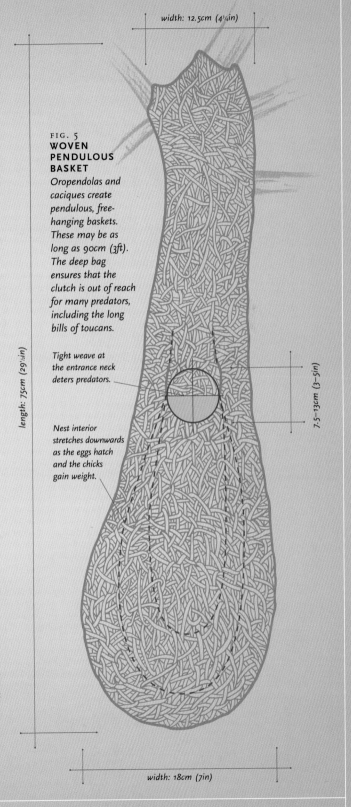

width: 12.5cm (4¾in)

FIG. 5
WOVEN PENDULOUS BASKET

Oropendolas and caciques create pendulous, free-hanging baskets. These may be as long as 90cm (3ft). The deep bag ensures that the clutch is out of reach for many predators, including the long bills of toucans.

length: 75cm (29½in)

7.5–13cm (3–5in)

Tight weave at the entrance neck deters predators.

Nest interior stretches downwards as the eggs hatch and the chicks gain weight.

width: 18cm (7in)

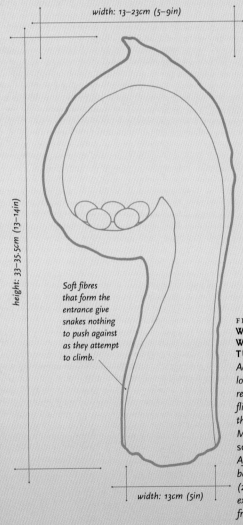

width: 13–23cm (5–9in)

height: 33–35.5cm (13–14in)

Soft fibres that form the entrance give snakes nothing to push against as they attempt to climb.

width: 13cm (5in)

FIG. 4
WOVEN RETORT WITH EXTENDED TUBE ENTRANCE

Access through a long entrance tube requires a vertical flight by the builder, the Red-vented Malimbe (Malimbus scutatus), from West Africa. The tube may be as long as 60m (2ft). This affords extra protection from predators.

FIG. 4 RED-VENTED MALIMBE NEST | FIG. 5 YELLOW-RUMPED CACIQUE NEST

MATERIALS AND FEATURES
Sooty-Capped Hermit Nest

The Sooty-capped Hermit (*Phaethornis augusti*) is in a hummingbird family of over 300 species. It hovers when feeding and building its nest. Hermit hummingbirds attach their nests beneath a living leaf, or similar site, in which the drooping shape forms a roof. Several hermits, such as the Green Hermit (*P. guy*) pictured below, attach a little cup of moss and other plant fibers to the leaf's tip to form their nest. The Sooty-capped Hermit's nest is even more extraordinary. First the female attaches a single, stout "cable" of spider silk to an overhead support. To the tail of this cable, she attaches moss and fragments of other vegetation by the "Velcro method" (see pages 76–77). More moss is added, in hovered flight, until the nest is formed beneath the suspension of a single attachment point on the rim of the leaf. The cup would tilt when the eggs and bird were in it, so to balance it the builder adds little lumps of mud to the nest's tail. These act as a counterweight to keep the nest level. In recent times, some Sooty-capped nests have been found hanging under a bridge, a highway culvert, or the ceiling of a building.

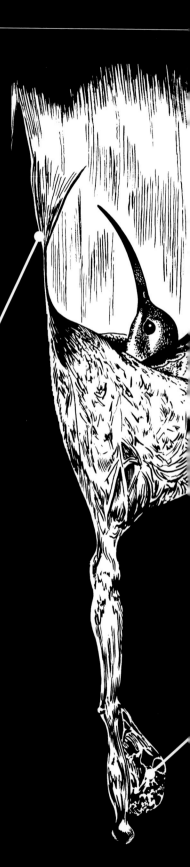

Spider silk threads
The use of silk to attach and construct hummingbird nests appears to be almost universal. Silk comes from spider webs and from cocoons. This Green Hermit in the Monteverde Cloud Forest Reserve, Costa Rica, has attached its nest to the tip of a palm leaf, and the silk thread suspensions are clearly visible.

Waterproofing

Protection from heavy rain is naturally an important consideration when constructing a nest in the rain forest. Large, drooping leaves not only offer a concealed site for the hermit's nest, they also form a convenient umbrella to fend off tropical rain. The nest of moss and silk also provides waterproofing, and the curved-over rim of the cup helps to prevent water running into the nest.

Counterweights

The Sooty-capped Hermit's nest employs ingenious counterweights to balance the structure, in the same way that these cranes employ a weight on the short arm to prevent the whole tower from toppling over when a load is picked up. The avian architect adds pellets of mud to the tail of the nest which add stability and prevent the nest from "capsizing" when occupied by the parent, eggs, or young.

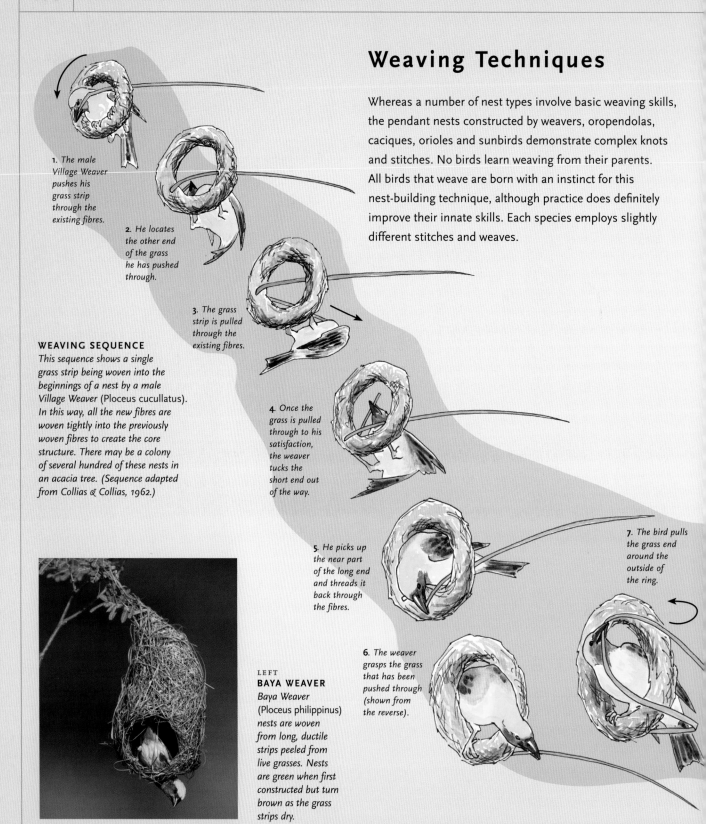

Weaving Techniques

Whereas a number of nest types involve basic weaving skills, the pendant nests constructed by weavers, oropendolas, caciques, orioles and sunbirds demonstrate complex knots and stitches. No birds learn weaving from their parents. All birds that weave are born with an instinct for this nest-building technique, although practice does definitely improve their innate skills. Each species employs slightly different stitches and weaves.

1. The male Village Weaver pushes his grass strip through the existing fibres.

2. He locates the other end of the grass he has pushed through.

3. The grass strip is pulled through the existing fibres.

WEAVING SEQUENCE

This sequence shows a single grass strip being woven into the beginnings of a nest by a male Village Weaver (Ploceus cucullatus). In this way, all the new fibres are woven tightly into the previously woven fibres to create the core structure. There may be a colony of several hundred of these nests in an acacia tree. (Sequence adapted from Collias & Collias, 1962.)

4. Once the grass is pulled through to his satisfaction, the weaver tucks the short end out of the way.

5. He picks up the near part of the long end and threads it back through the fibres.

7. The bird pulls the grass end around the outside of the ring.

6. The weaver grasps the grass that has been pushed through (shown from the reverse).

LEFT
BAYA WEAVER
Baya Weaver (Ploceus philippinus) nests are woven from long, ductile strips peeled from live grasses. Nests are green when first constructed but turn brown as the grass strips dry.

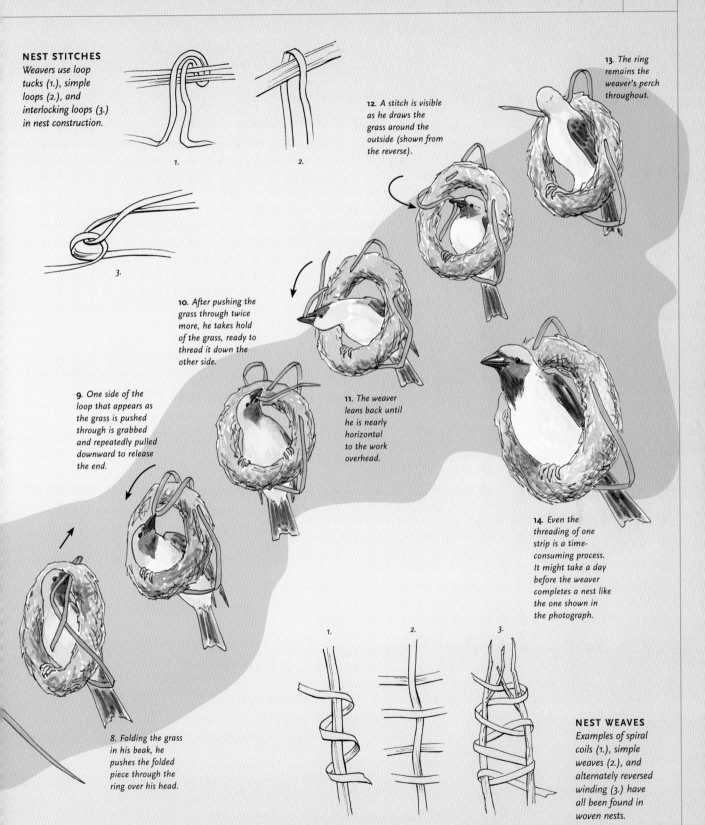

NEST STITCHES
Weavers use loop tucks (1.), simple loops (2.), and interlocking loops (3.) in nest construction.

1.

2.

3.

12. A stitch is visible as he draws the grass around the outside (shown from the reverse).

13. The ring remains the weaver's perch throughout.

10. After pushing the grass through twice more, he takes hold of the grass, ready to thread it down the other side.

9. One side of the loop that appears as the grass is pushed through is grabbed and repeatedly pulled downward to release the end.

11. The weaver leans back until he is nearly horizontal to the work overhead.

14. Even the threading of one strip is a time-consuming process. It might take a day before the weaver completes a nest like the one shown in the photograph.

8. Folding the grass in his beak, he pushes the folded piece through the ring over his head.

1.

2.

3.

NEST WEAVES
Examples of spiral coils (1.), simple weaves (2.), and alternately reversed winding (3.) have all been found in woven nests.

CASE STUDY
Baltimore Oriole

The Baltimore Oriole is a member of the icterid family and a summer visitor to the eastern and central United States and to southern Canada as far west as Alberta. The Baltimore Oriole is replaced in the western states by Bullock's Oriole (*Icterus bullocki*). The oriole constructs an elaborate and durable pensile nest using tens of thousands of stitches and rapid shuttle movements.

Nest site and habitat

These two orioles are closely related and both breed in areas of scattered trees, even in towns, orchards and open deciduous woodland. Males usually arrive first, and once a pair has mated the female selects a site most often about 6–9m (20–30ft) up in a tree or shrub, where the nest can be suspended from the fork of a twig near the end of a branch.

Nest materials

The female builds, but is followed everywhere protectively by the male, who sometimes transports nest material. The female gathers long plant fibres, strips of vine or tree bark, hair, and even string or yarn. This becomes the nest's shell, which is lined with finer grasses, moss, hair and plant down.

Nest building

The nest is a pensile structure bound by fibres to the tips of the forked twig. Depending on the site, there may be more than the two or three basic attachment points, which become the nest's rim. The purse-shaped nest grows downward from these points as more and more fibres are woven in.

Movements and stitches

The finished nest is 7.5–10cm (3–4in) across. Herrick believed the one he studied for 4½ days from start to finish contained about 200 slender fibres. He calculated that a typical nest might contain 10,000 stitches and thousands of knots and loops. To complete the nest, the dizzying number of 20,000 shuttle movements of head and bill would have been required.

Classification

ORDER	Passeriformes
FAMILY	Icteridae
SPECIES	*Icterus galbula*
RELATED SPECIES	Other New World orioles
NEST TYPE	Hanging, woven
SPECIES WITH SIMILAR NESTS	Other icterids of the New World, Old World orioles (family Oriolidae), Australian honeyeaters
NEST SPECIALIZATION	Suspension from twig ends

LEFT
BALTIMORE ORIOLE NEST
This female Baltimore Oriole is busy firming up the nest's rim. The photograph shows this species' preference for light-coloured fibres, in contrast to the Orchard Oriole (Icterus spurius), which often uses green grass for its flexibility, thus making the Baltimore's nest more conspicuous in a leafy tree.

BELOW
NEST CONSTRUCTION
This sequence shows the intricate nature of the delicate, growing structure (after Herrick, 1911).

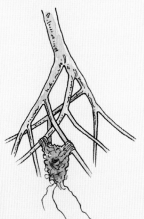
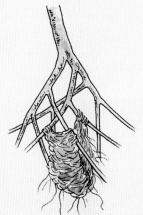
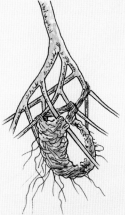
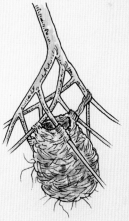
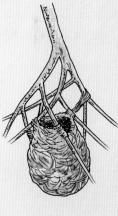

1. *The nest starts with two primary attachment points.*

2. *After around three hours, there is a third attachment point, and many threads hanging.*

3. *The rim is finished after 10 hours' work, and the nest wall is blocked out except for one side. The streamers are later worked into the nest's shell.*

4. *On the third day, after about 20 hours' work, the nest is completely outlined, but the walls, especially the one opposite the first completed side, is thin. The nest is now secured at six points.*

5. *The walls are lined and no longer thin, yet they are flexible and can expand as the brood of 4–5 chicks grows. The nest has been completed in 4½ days.*

CASE STUDY
Goldcrest

The Goldcrest, Europe's smallest bird, provides an example of a delicate nest formed with threads of spider silk. Resident across much of Europe and Asia to about 100°E, and in isolated populations in the Far East, this bird is often found in well-grown conifers, especially in the breeding season. Because of its northern breeding distribution, it shelters its nest from the prevailing wind and insulates it with feathers.

Nest site

Most nests are positioned for shelter from the prevailing wind, suspended near the end of a horizontal conifer branch. The beautiful, well-camouflaged little cradle is built so tightly against the foliage above that it is hard for humans to spot. Most nests are built between 2m (6½ft) and 4m (13ft) high, but some have been found as high as 12m (39ft).

Nest and materials

The nest is an almost spherical cup of moss and lichen, with a small entrance at the top over the rim. The entrance is often restricted by the pine and fir needles and the closeness of the twigs above. The moss and lichen of the cup are the 'bricks' of the construction and they are held together with the 'mortar' of silk from spider webs. As shown above right, a framework of moss and silk is woven around several twigs, from which a three-layered cup is suspended.

Nest insulation

The walls constructed by the Goldcrest are comparatively thick for a little nest, reducing its outer diameter of 7.5–9cm (3–3½in) to a cup of only 3.75cm (1½in). The narrow opening also helps to maintain warmth. Insulation is a key feature of this construction, and the lining may include a very high number of feathers. In 1969, von Haartman counted 2,672 feathers in a single nest. The Goldcrest may also include hair in the lining. The birds have a clutch size of 7–10 eggs, sometimes more.

Classification

ORDER	Passeriformes
FAMILY	Regulidae
SPECIES	*Regulus regulus*
RELATED SPECIES	Kinglets, Old World 'leaf warblers'
NEST TYPE	Hanging, stitched
SPECIES WITH SIMILAR NESTS	Orioles (both families), vireos
NEST SPECIALIZATION	Silk stitching, feather insulation

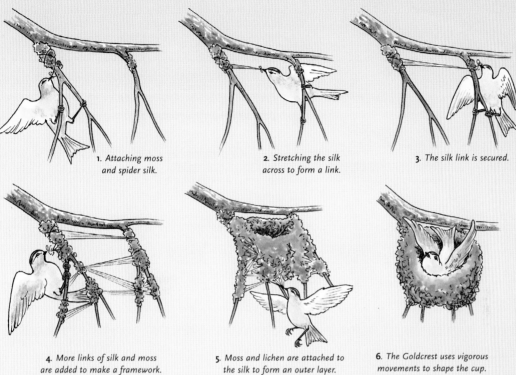

1. Attaching moss and spider silk.

2. Stretching the silk across to form a link.

3. The silk link is secured.

4. More links of silk and moss are added to make a framework.

5. Moss and lichen are attached to the silk to form an outer layer.

6. The Goldcrest uses vigorous movements to shape the cup.

LEFT
GOLDCREST NEST CONSTRUCTION
First, the bird attaches spider silk and moss to a twig (1.). The silk is stretched to form a link with a second twig (2.) and secured (3.). Further silk links are stretched across and stitched to pads of moss to form a framework. Then a three-layered cup is stitched together. An outer layer (5.) is made by attaching moss and lichen to the silk. A middle layer of moss and lichen thickens the walls. A layer of feathers and hair forms the lining. Vigorous movements of breast, feet and wings are used to shape the cup (6.).

LEFT
SHELTERED GOLDCREST NEST
A finished nest sits very close to the underside of a conifer branch. The tiny bird, some 8.5–9.5cm (3⅓–3¾in) long, only just fits in the space to feed the young. The nest is bound around three main sprays of needles, which are felted into the cup.

CASE STUDY
Common Tailorbird

The Common Tailorbird stitches with silk to construct its nest in the cone formed between the two drawn-together sides of a leaf or between two or more leaves — one of the most remarkable constructions in the bird world. These sedentary birds are found across India and Sri Lanka, southern China and Java. They are commonly found in orchards, scrub and gardens on the outskirts of villages and towns.

Nest site
The principal requirement of a suitable nest site is a shrub or tree with large, broad, spearhead-shaped and drooping leaves that are strong enough to make a safe cradle for the nest. A mango tree makes a good choice. The chosen leaf will be only a few feet above the ground. The weight of the nest draws the top of the leaf over the nest to shelter it.

Nest building
The female does most of the work, while the male defends the territory. First, the female brings the two edges of the leaf together by twisting strands of spider silk around them. She then pierces a hole in the edges of the leaf near the tip with her beak, pushes in a silk thread, pulls the thread through, then punches it through again. The leaf's tension draws the silk into a knot. The tailorbird repeats this up the leaf's edges, with several failures where the leaf splits or the thread breaks. Mike Hansell has identified lepidopterous silk, spider silk and plant down in the thread. The stitching of the leaf may take two days, then the nest is made within the security of the shaped leaf. The male transports nest material of vegetable cotton, down, fine grass and hair for the lining. If the nest rim rises above the top of the stitched cone, it is camouflaged with green vegetation attached with silk. Extra stitches may be added from the leaf into the nest wall.

Eggs and young
The female lays 2–5 eggs, which hatch after 12 days. Both parents feed and raise the chicks, which fledge at about 24 days.

Classification

ORDER	Passeriformes
FAMILY	Cisticolidae
SPECIES	*Orthotomus sutorius*
RELATED SPECIES	Other Old World warblers
NEST TYPE	Hanging, stitched
SIMILAR SPECIES WITH NESTS	Other tailorbirds, related warblers, spider hunters.
NEST SPECIALIZATION	Leaf shaped with silk stitches

CASE STUDY

Brown-backed Honeyeater

Honeyeaters form the most successful family of birds in Australia, where nearly 70 species are found. It is resident in tea-tree or paperbark (*Melaleuca* spp.) swamps, mangroves and woodland near permanent water in northeast Queensland and New Guinea. This family is characterized by a brush-tipped tongue for feeding on nectar. The Brown-backed Honeyeater constructs its hanging nest at the tip of a twig.

Nest site

The Brown-backed Honeyeater usually suspends its nest up to 2.4m (8ft) above the ground and sometimes as high as 4–5m (13–16ft). Often, the nest hangs at that height over water in the mangroves or swamp.

Nest building

To begin construction of its nest, the honeyeater first twists strands of *Melaleuca* bark around the tip of a thin, drooping twig. Gradually, a somewhat untidy dome of more strips and pieces of bark and strands of grass is formed, bound with spider silk. The dome's edges are added to until the bird has constructed a bulky cup, with a largish, round entrance on one side. The nest is lined with small pieces of bark. The result is a hooded, cupped, hanging nest that offers good protection from both predators and the elements.

Eggs and young

The main breeding season of the Brown-backed Honeyeater extends from August to February. The female lays 2–3 white, sparingly marked eggs. Incubation by the female takes about two weeks, and both adults feed the chicks for a further two weeks.

Classification

ORDER	Passeriformes
FAMILY	Meliphagidae
SPECIES	*Ramsayornis modestus*
RELATED SPECIES	Nearly 170 other honeyeaters
NEST TYPE	Hanging
SPECIES WITH SIMILAR NESTS	Some sunbirds and warblers
NEST SPECIALIZATION	Grass and bark stitched with silk

CHAPTER NINE

Mound Nests

Mound nests are built on the ground, and this style of avian architecture ranges from simple piles and pits to some of the largest constructions of the bird world. This chapter deals with two distinct types of mound architecture. The first is the 'incubator mound', in which eggs are buried and their temperature regulated, and the chicks are born more precocially developed than any other kind of bird. The second, which shares features with platform, mud and aquatic forms, is the 'mound nest', where materials are piled up to form a more typical protection for eggs and chicks, and the eggs are placed in a depression on the top.

The true mound builders are the architects of the uniquely engineered incubator mounds. These birds are the megapodes – literally 'big feet'. The 22 members of the family Megapodiidae are divided into three ecological groups: scrub fowl, brushturkeys and the Malleefowl (*Leipoa ocellata*), and all are found only in Australasia. Several species can construct enormous mounds of soil and vegetation in which the eggs are laid. The vegetation generates heat as it rots, incubating the eggs. The male controls the mound's temperature, checking it with his beak and using his strong feet to add or remove leaf litter. Megapodes' eggs are unique in having large yolks that are up to 50–70 per cent of the egg's weight. The chicks develop in the mound and hatch completely feathered, and after digging themselves out of the mound are able to run, feed and even fly on the same day. The adults take no further part in their upbringing.

The other mound builders included in this chapter scrape or pile up mud, rocks and other materials to create solid mound architecture that elevates nest and eggs above ground level. The Light-mantled Albatross (*Phoebetria palpebrata*), for example, pairs for life, and nests in loose colonies on islands in the Southern Ocean. It builds a solid mound of peat and mud 16–31cm (6¼–12½in) high and 45–55cm (17¾–21½in) round at the base, on top of which a slight depression is lined with grass to take the single egg.

Not all mound builders use such soft materials; the Black Wheatear (*Oenanthe leucura*) and Adélie Penguin (*Pygoscelis adeliae*) use stones to form a substantial part of their nests. The Horned Coot (*Fulica cornuta*) builds an artificial island from astonishing numbers of stones, with a cup nest of vegetation to hold the eggs. Adélie Penguins also construct mounds of stones in colonies where strong competition for materials encourages them to steal from their neighbours in a bid to construct a protective nest.

RIGHT
**AMERICAN FLAMINGO
MOUND NESTS**
Part of a colony of American Flamingos (Phoenicopterus ruber) *at Inagua National Park, in the Bahamas, with the eggs clearly visible on their raised-up mound nests of mud.*

BLUEPRINTS
Mound Nest Structures

Scrub fowl, brushturkeys and Malleefowl (*Leipoa ocellata*) make incubator mounds in which eggs are buried. The characteristic blueprint is a ditch surrounding a mound of material that generates heat like a compost heap as it rots. But some species simply bury their eggs in sand. The other mound blueprint, for species that pile up mounds of mud, stones and plant materials to create elevated nests for eggs and chicks, includes the Mute Swan (*Cygnus olor*), flamingos, cormorants, some albatrosses and the Adélie Penguin (*Pygoscelis adeliae*).

FIG. I
INCUBATOR MOUND

The Malleefowl (Leipoa ocellata) constructs a huge pit and mound by scraping up earth and leaf litter with its strong legs and feet. Eggs are laid in a hollow and buried in the leaf litter, which is covered with sandy soil. The rotting leaf litter works like a compost pile and its heat incubates the eggs. The male controls the mound's temperature by adding or removing the leaf litter to manage incubation.

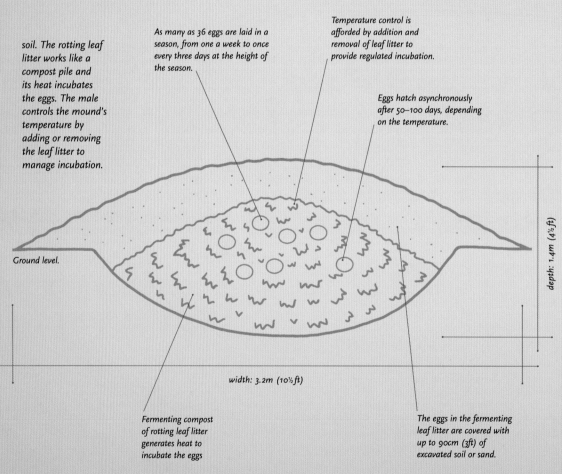

As many as 36 eggs are laid in a season, from one a week to once every three days at the height of the season.

Temperature control is afforded by addition and removal of leaf litter to provide regulated incubation.

Eggs hatch asynchronously after 50–100 days, depending on the temperature.

Ground level.

depth: 1.4m (4½ ft)

width: 3.2m (10½ ft)

Fermenting compost of rotting leaf litter generates heat to incubate the eggs

The eggs in the fermenting leaf litter are covered with up to 90cm (3ft) of excavated soil or sand.

FIG. I MALLEEFOWL NEST

VARIETIES OF STRUCTURE

The mounds of Malleefowl have caused the name 'mound builders' to be applied to the whole family. However, the Maleo (*Macrocephalon maleo*) of the Indonesian island of Sulawesi lays its eggs in a hole in volcanic sand on a beach instead of constructing a mound. It has weaker feet than the Malleefowl, so is not such an adept digger. These birds, and scrubfowl (*Megapodius* spp.) on other islands, return to the forest as soon as they have covered the eggs.

FIG. 2
MOUND AND TRENCH

One of the largest of all seabirds, the Wandering Albatross (Diomedea exulans) comes to land only to breed on one of several islands in the Southern Ocean. The nesting cycle lasts 13 months, so they breed only every other year. They nest in scattered colonies, with nests spaced about 20–45m (20–50yd) apart. They scratch a circular trench, piling up a mound of soil, moss and grass high enough to elevate the nest above snow.

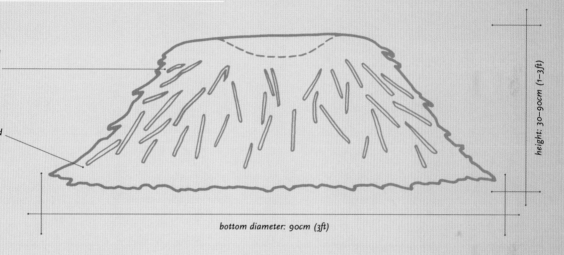

Mound of moss, soil and grass is built up and trampled down until the albatross nest resembles a small volcano.

Circular trench is scraped out by the male's beak, with the generated mound material piled up in the centre.

height: 30–90cm (1–3ft)

bottom diameter: 90cm (3ft)

FIG. 3
MUD MOUND

Flamingos are very gregarious, often nesting together in their thousands. The American Flamingo (Phoenicopterus ruber) makes a typical mud mound, which is very similar to those of the Greater Flamingo (P. roseus) in Europe, India and Kenya. Both sexes share the building, standing in the middle of the chosen spot on damp, muddy ground. They scrape up mud pellets within reach from the surrounding mud and roll them into position and up the sides of the mound. Building takes several days of sporadic work. A conical mound is produced with a shallow cup on top, surrounded by a circular trench, which may be as deep as 20cm (8in). Nests built on sand are very much smaller. Most nests will be 20–50cm (8–20in) apart. The nest site is changed when conditions are not right, even if breeding was successful before.

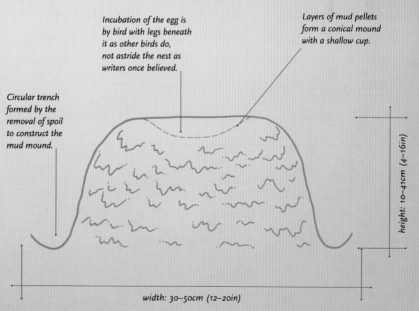

Incubation of the egg is by bird with legs beneath it as other birds do, not astride the nest as writers once believed.

Layers of mud pellets form a conical mound with a shallow cup.

Circular trench formed by the removal of spoil to construct the mud mound.

height: 10–41cm (4–16in)

width: 30–50cm (12–20in)

FIG. 2 WANDERING ALBATROSS NEST | FIG. 3 AMERICAN FLAMINGO NEST

MATERIALS AND FEATURES
Black Wheatear Nest

The Black Wheatear (*Oenanthe leucura*) is a rock-loving species that builds its nest from late March onwards in a hole in a rock face or wall, at ground level or at a height of up to 4.6m (15ft). First, the male chooses a nest site, then he carries stones in his bill to the site and performs a song-flight or display. Some stones fall, and the sound and the display appear to stimulate a female to select the male and his nest site. Male and female then carry one stone at a time to make a wall or mound at the entrance to the nest. The ornithologist Frank Richardson found a few stones or as many as several hundred; 358 were counted at one nest, with each on average weighing 6–7g (1/4oz); many weighed much more. One recorded stone weighed an amazing 28g (1oz) – carried by a 38-g (1 1/3-oz) bird. The nest is constructed on the stone foundations from materials including grasses, plant stems and leaves, lined with finer grasses, wool and feathers.

Stone foundations
In this rock crevice in Spain, the nest's lip shows clearly above the foundations of stones. The stones have been variously described as a foundation for the nest; a wall to prevent the nest sliding; protection from bad weather and predators; and a part of courtship behaviour. In 22 of 37 nests studied in southeastern Spain, the stones served the purpose of either creating a cup for the nest or holding it securely on a sloping surface.

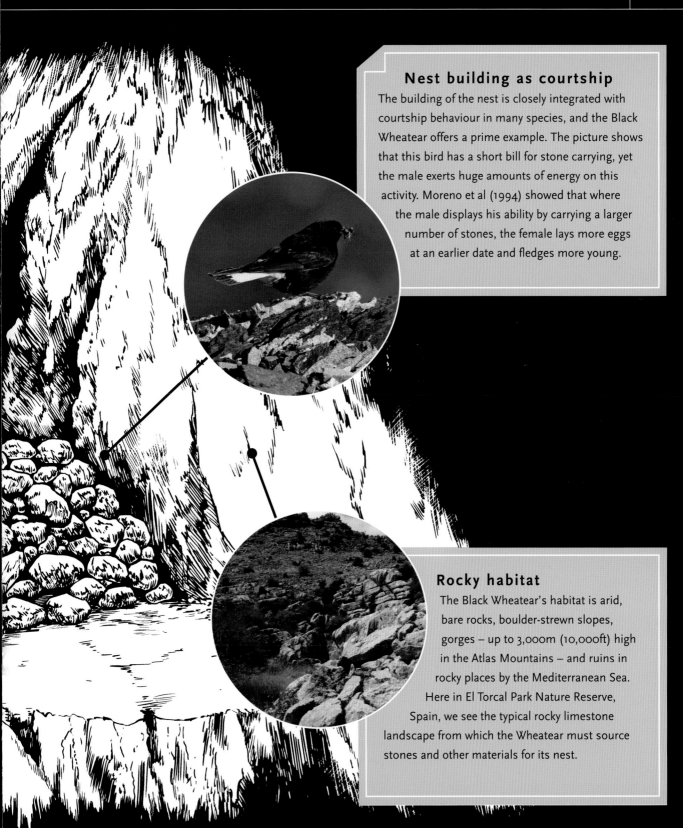

Nest building as courtship

The building of the nest is closely integrated with courtship behaviour in many species, and the Black Wheatear offers a prime example. The picture shows that this bird has a short bill for stone carrying, yet the male exerts huge amounts of energy on this activity. Moreno et al (1994) showed that where the male displays his ability by carrying a larger number of stones, the female lays more eggs at an earlier date and fledges more young.

Rocky habitat

The Black Wheatear's habitat is arid, bare rocks, boulder-strewn slopes, gorges – up to 3,000m (10,000ft) high in the Atlas Mountains – and ruins in rocky places by the Mediterranean Sea. Here in El Torcal Park Nature Reserve, Spain, we see the typical rocky limestone landscape from which the Wheatear must source stones and other materials for its nest.

Malleefowl Incubating Technique

The Malleefowl (*Leipoa ocellata*) of semiarid southern Australia provides a prime example of how the megapodes use an incubator mound to control the temperature of their buried eggs. The breeding pair between them work at the mound for 11 months of the year. This series of drawings follows the work of the Malleefowl pair through the breeding season as they strive to maintain a constant temperature of 34°C (93°F) in the mound. The fluctuations in temperature during 24 hours and during the year mean that the bird has to work hard to provide a constant incubation temperature for the breeding season. By summer, there may be a day-to-night air-temperature fluctuation of 16°C (29°F). The female may lay 35 eggs over a long period of time, digging a new pit each time another egg is laid.

1. WINTER
During the winter, before the rains, the Malleefowl pair dig a large pit about 90cm (3ft) deep and 5m (16½ft) in diameter, piling the spoil around the rim.

2. LATE WINTER
In late winter they fill the pit with vegetation. The rains come, the vegetation rots and the temperature at the bottom of the pit may rise to 60°C (140°F).

3. WINTER / SPRING
By early spring, the birds sense the rise in temperature and cover the fermenting material with sand.

LEFT
MALE MALLEEFOWL CONSTRUCTING
This male Malleefowl is in the act of sending a shower of nest material up the mound's slope behind him. He scrapes soil, sand and vegetable debris onto the mound from within a radius of 50m (164ft).

5. SPRING / SUMMER

A new pit is dug each time another egg is laid. The birds check the temperature by pushing their beaks into the mound – in this way the male discovers whether he needs to open up the mound to cool it, or cover it more to build up the temperature. Day-to-night temperature fluctuations in the summer are highly demanding.

6. AUTUMN / WINTER

The mound is cleared out in autumn and redug in the winter.

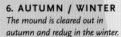

4. SPRING

A central chamber of sand and leaves is left, in which the female lays the first egg. The female tests the nest before she lays a new egg. She needs to dig past the surface, which is only 15–19°C (59–66°F), and not go too deep where temperatures can rise to 40–60°C (104–140°F). The ideal temperature zone is 34°C (93°F).

CASE STUDY
Australian Brushturkey

The Australian Brushturkey provides a good example of birds who bury their eggs to incubate them. It has a wingspan of 85cm (33½in), and is confined to rain forest and wet open forest with good ground cover in New South Wales and Queensland, Australia. Brushturkeys are communal birds; a dominant male has several females and one or more younger males in his group. Once hatched, chicks feed themselves.

Incubator mound
The dominant male digs and scratches a mound of leaves and other plant material until it is 1.5m (5ft) high and 4m (13ft) across. This is added to and repaired each year until the mound may be 1.8m (6ft) high and 9m (30ft) in diameter.

Temperature control
A hole 45–61cm (1½–2ft) deep is scratched in the top, and 16–30 or more eggs are laid there, large end uppermost, at intervals of several days. Often, more than one female will lay eggs in the nest of a dominant male. Each egg is carefully covered over with nest material. The nest works like a huge compost pile. The male checks the temperature by sticking his beak into the mound, and adds or removes material with his strong feet to maintain a temperature of 33–35°C (91–95°F).

Precocial young
After about seven weeks, the chicks hatch one by one and dig themselves out of the mound. They are completely feathered and able to run. The chicks receive no parental care, but soon learn to recognize other young of a similar age and band together. Scientists believe that other birds pecking the ground for food and their colour are factors in helping recognition. They have also discovered that cold weather at nesting time results in more males and warm weather in more females. A mound kept at about 34°C (93°F) results in an equal sex ratio, but above or below, the imbalance occurs – this is the first time this has ever been shown for birds, although it is known to occur in some reptiles.

Classification

ORDER	Galliformes
FAMILY	Megapodiidae
SPECIES	*Alectura lathami*
RELATED SPECIES	Megapodes, curassows, pheasants
NEST TYPE	Incubator mound
SPECIES WITH SIMILAR NESTS	Only other megapodes
NEST SPECIALIZATION	Temperature control

1. *The male builds an incubator mound up to 1.5m (5ft) high.*

2. *Chicks hatch at intervals from the top.*

LEFT
AUSTRALIAN BRUSHTURKEY INCUBATION
The male builds using leaves and other plant material to create a mound up to 1.5m (5ft) high (1.). The birds usually nest only once a year from August to January. An egg is laid every 3–5 days. One female may lay 15 in a season. If the male has a harem, there may be 30 or more eggs in the mound at one time. Eggs start to incubate as soon as they are covered, so the young hatch one by one, digging themselves out of the top of the mound at intervals (2.).

LEFT
PRECOCIAL CHICK
The precocial chick takes about a day to dig its way out of the mound. It is completely feathered, and once out of the nest it can feed, run and in a few hours is even able to fly, probably never seeing its parents. Waders and wildfowl have chicks that can feed and run, but they are covered in down not feathers and are cared for by the parents.

CASE STUDY
Adélie Penguin

The Adélie Penguin builds a mound nest of stones, and exhibits a fascinating stone-stealing behaviour. This is the most common penguin and the smallest Antarctic species. They breed in the Southern Hemisphere spring, coming ashore from October onwards, having wintered on the sea ice. Experienced birds arrive first and may walk 20–40km (12–25 miles) across the ice before they reach land.

Nest site

Some birds at Hope Bay colony have up to 320km (200 miles) to walk to their traditional nest site – right to the spot where they nested the previous year. Adélie Penguins prefer to build their simple mound nests on exposed, well-drained sites with plenty of stones. Exposed sites are less likely to have snowdrifts, which could bury the birds and eggs.

Nest building

The pair go through a series of display rituals culminating in the creation of a mound, which they make by lying on their bellies and scraping with their feet. Stones are added until the cup is raised well clear of ground level – this prevents the eggs from becoming chilled if there is flooding from melted snow. Stones are always in short supply, so stealing occurs frequently. These thefts do have the benefit of spreading stones through the colony. Nests have been described as 'two pecking distances apart', which helps to reduce aggression between penguins. The colony has the advantage of encouraging synchronized breeding in the short season, providing safety in numbers. The disadvantage is that the colony is conspicuous and attracts South Polar Skuas (*Catharacta maccormicki*), the Adélies' only predator on land.

Eggs and young

Incubation of the two eggs is shared by the parents for 30–40 days; chicks are guarded for another 20 days or more, after which they form crèches of many hundreds. Adults returning with food find their chick by recognizing its call, even in the hubbub of sound.

Classification

ORDER	Sphenisciformes
FAMILY	Spheniscidae
SPECIES	*Pygoscelis adeliae*
RELATED SPECIES	Only nearly 20 other penguins
NEST TYPE	Mound
SPECIES WITH SIMILAR NESTS	Very few
NEST SPECIALIZATION	Stone mounds in colonies

1. The approach.

2. The theft.

3. The escape.

4. The thief presents the prize to its mate.

LEFT
ADÉLIE PENGUIN STONE STEALING
The Adélie Penguin pair collect enough stones to create a low mound, which helps to protect the two eggs in the centre from melting snow. Each time one of the pair greets the other, it offers a stone as a gift. The stones are important in the birds' breeding cycle, but there are often insufficient materials to fulfil the needs of a huge colony. Very often this means that the returning bird will go up to a nest en route, snatch a stone, then run away to its own nest and offer it to its mate, who will add it to the pile.

LEFT
ADÉLIE PENGUIN INCUBATION
Incubation starts in December, during the short Antarctic summer when the temperature is still only −2°C (28°F). The penguin's mate goes away for several days to feed, during which time the incubating bird does not eat. Eventually, it will be relieved by the incoming bird and be able to feed. This sharing of duty goes on until March, when the families all go to the ice floes until October or November. They will come ashore for the next breeding season.

CASE STUDY

Horned Coot

The Horned Coot is restricted to the altiplano lakes of Bolivia, Argentina and Chile, in the high Andes some 3,000–5,200m (9,800–17,000ft) above sea level. The climate is cool and mostly dry with a rainy season from December to March, with the coots breeding at the beginning of this period. The Horned Coot builds an extraordinary mound nest that involves constructing an island of rocks.

Nest foundations

Horned Coots build their nest about 4m (13ft) from the shore in the shallow water of a lake. They pile up stones offshore to form an artificial island. The pair collect the stones from the shore and carry them one by one to the building site. Hundreds of trips are needed to build a mound, which has been estimated to weigh as much as 1.3 tonnes. This develops into a huge cone as wide as 4m (13ft) across and 60–90cm (2–3ft) high. The top is just above the surface of the water.

Nest and eggs

The pair then dive into the water to collect algae and aquatic vegetation to build a nest on top of the island. This forms a truncated cone on the stones that rises 35.5–60cm (14–24in) above the water and provides a cup for the eggs about 60cm (2ft) across. A clutch of 3–5 eggs is laid, which remains safe because there is little or no fluctuation in the water level. The mound can be reused the following year.

Threatened habitat

The birds' journeys to and from the nest and the shore become even more of an effort in competition with other pairs doing the same. The species can bear the increased cost to nest building, as long as its habitat is safe. However, BirdLife International lists the Horned Coot as Near Threatened, partly due to the contamination of lakes. It also suffers from hunting, egg harvesting, and some predation by Andean Gulls (*Larus serranus*).

Classification

ORDER	Gruiformes
FAMILY	Rallidae
SPECIES	*Fulica cornuta*
RELATED SPECIES	Crakes, rails, moorhens
NEST TYPE	Mound
SPECIES WITH SIMILAR NESTS	None
NEST SPECIALIZATION	Island of stones

LEFT
HORNED COOT MOUND NEST

The Horned Coot is the second largest coot – 45–60cm (1½–2ft) long, compared with the American Coot's 39cm (15½in). It is one of the world's rarest birds in addition to having one of the strangest nesting habits. The population is scattered among the mountain lakes. This one was at Laguna Miscanti-Miñiques, Chile, where the total population is around 600. The nest mound looks more like a natural rock island in the water.

The uppermost part of the nest.

A cross-section view reveals the true size of the nest.

LEFT
ISLAND MOUND

Only a drawing can really show the nest's extraordinary position and structure. Grebes' and Common Coots' nests, as we have seen (see pages 46–55), are aquatic but attached to or very close to vegetation. This solid offshore island is completely built and maintained by the two birds in arid mountainous conditions.

CHAPTER TEN

Colonies & Group Nests

About one-tenth of the world's bird species are colonial, constructing their nests in close proximity to their neighbours. Colonies and group nests include many of the architectural types featured on previous pages, and this chapter is concerned less with the design of individual nests. It looks instead at the behaviour of communal nest building, and at the density and positioning of colonial nests.

The study of birds that breed together has shown there to be three kinds of social organization. In colonial groups, which include penguins, auks, gulls, terns, swifts and martins, many birds breed together but do not give each other any direct help. We might equate their many individual nests, built in close proximity to each other, to a densely populated human city.

In cooperative groups, a few species, such as the Sociable Weaver (*Philetairus socius*), live on intimate terms and assist in building one huge nest in which each pair has a separate compartment. This shared nest might be likened to a human apartment block.

The third kind of social organization is the communal group, in which passerines live in a highly organized community – the Florida Scrub Jay (*Aphelocoma coerulescens*) is one example. A few adults build a nest, one or two females lay eggs, and all the group feed the nestlings.

For colonial nesting to work, there must be minimal competition for food. Seabirds can be colonial because their feeding grounds may be a long distance away. For Rooks (*Corvus frugilegus*), food sources are unpredictable and cannot be defended so there is no disadvantage in sharing the same territory.

The advantages of colonial nesting include protection from predators – there are several pairs of eyes to spot the hunter, as shown by the photograph above of a colony of Atlantic Puffins (*Fratercula arctica*), and birds can act in collective defence. Colonial birds have the advantage of being able to see where other birds in the colony go for food.

The densely packed area of colonial nest territory brings some disadvantages. The close grouping of nests makes them obvious to a predator. Shortage of nest sites can also present a difficulty, as can be seen at those seabird colonies which are confined to a finite series of cliff ledges. There is also the matter of aggression between neighbours: some species, such as Rooks, steal each others' nest material, while males regularly have to defend their mates from the sexual advances of neighbours.

RIGHT
BLACK-BROWED ALBATROSS NESTS
These Black-browed Albatrosses (Thalassarche melanophris) *on Steeple Jason Island, the Falkland Islands, nest on average only 1.5m (5ft) apart. There were over 171,000 pairs there in the last census in 2005.*

Colonial Nest Structures

Colonial nest blueprints are characterized by groups of nesters, including auks, albatrosses, herons, terns and some swallows, constructing their individual nests in close proximity to each other. Although neighbouring nests are close, they tend to be regularly spaced, as in a cliff-top gannetry. In a Brown Noddy (*Anous stolidus*) ternery, the spacing may be more generous, due to the availability of nest sites in the trees.

FIG. I
COLONIAL SCRAPE NEST
Common Terns (Sterna hirundo) nest in colonies on coasts and by lakes across the Northern Hemisphere, and winter in the far south. These birds are gregarious throughout the year. In a colony of 55 nests in Virginia, USA, the average distance between each was 1.6m (5ft 3in). Colonies generally range in size from a few to several hundred, but thousands have been recorded in the Netherlands.

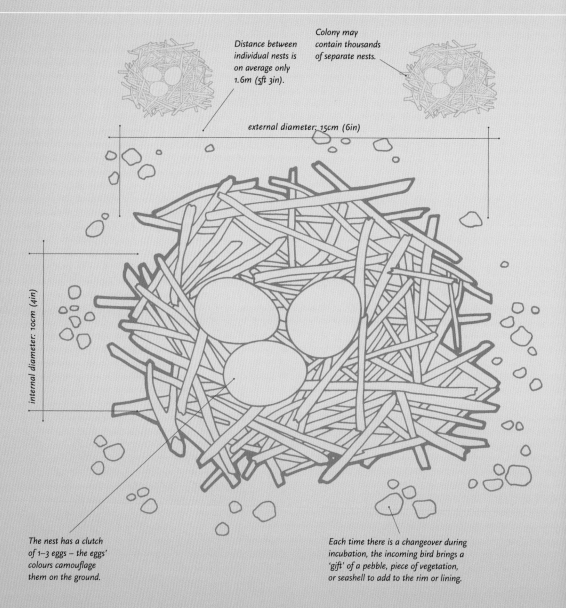

Distance between individual nests is on average only 1.6m (5ft 3in).

Colony may contain thousands of separate nests.

external diameter: 15cm (6in)

internal diameter: 10cm (4in)

The nest has a clutch of 1–3 eggs – the eggs' colours camouflage them on the ground.

Each time there is a changeover during incubation, the incoming bird brings a 'gift' of a pebble, piece of vegetation, or seashell to add to the rim or lining.

FIG. I COMMON TERN NEST

VARIETIES OF STRUCTURE

A wide variety of birds nest in groups, and therefore their nests vary greatly both in type and construction materials. Some, such as the Common Tern (*Sterna hirundo*), build a basic scrape nest with a lining of vegetation; the Lesser Flamingo (*Phoeniconaias minor*) creates a mound from the surrounding mud; Rooks (*Corvus frugilegus*) build their colony of cup nests, known as a rookery, out of sticks.

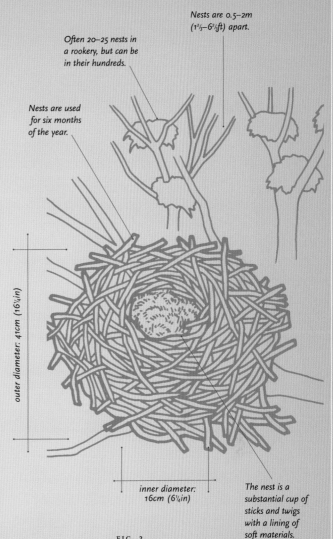

Nests are 0.5–2m (1²⁄₃–6²⁄₃ft) apart.

Often 20–25 nests in a rookery, but can be in their hundreds.

Nests are used for six months of the year.

outer diameter: 41cm (16¹⁄₄in)

FIG. 2
COLONIAL MOUND NEST

For the last 45 years Lake Natron in Tanzania has been East Africa's only breeding site of the Lesser Flamingo. This very caustic soda-lake gives the 1.5–2.5 million birds protection from predators and an excellent food supply. The nest is built out of mud dug from around the mound. The site and the species are threatened by industrial development.

Cluster of 5 nests per 1m² (10ft²). Distance between individual nests can be as little as 50cm (19³⁄₄in).

The top of the nest is never as hot as the extremely high temperature of the surrounding mud, so chicks stay close to or on the nest mound for several days.

Where mud has hard coating of soda, nests are arranged linearly along the cracks.

inner diameter: 16cm (6¹⁄₄in)

The nest is a substantial cup of sticks and twigs with a lining of soft materials.

FIG. 3
COLONIAL CUP NEST

The Rook is a member of the crow family – the only species of crow to nest colonially. The rookery is clustered in treetops. A 'colony' may be a complex of several groups, in each of which are usually several nests in every tree. The nest does not last the winter as well as that of its close relative, the Carrion Crow (Corvus corone). Nests that survive may be reclaimed by the previous owners and repaired, or newcomers will salvage the remaining sticks to build their own nest.

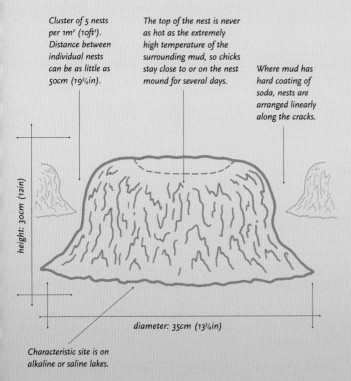

height: 30cm (12in)

diameter: 35cm (13³⁄₄in)

Characteristic site is on alkaline or saline lakes.

FIG. 2 LESSER FLAMINGO NEST | FIG. 3 ROOK NEST

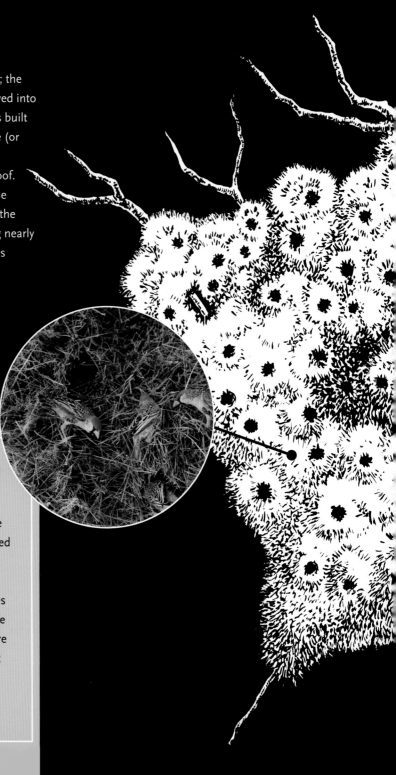

MATERIALS AND FEATURES

MATERIALS AND FEATURES
Sociable Weaver Nest

The Sociable Weaver (*Philetairus socius*) is well named; the
bird is sociable throughout the year. Its nest has evolved into
one of the largest structures built by birds. The nest is built
from the top down. In the green of the crown of a tree (or
sometimes on a telephone pole), two or three pairs
construct a platform of twigs and grasses to form a roof.
It has a sloping thatch so that rain flows away from the
entrances. The birds work a thick mat of grasses into the
underside. Straw stems are pushed into this, pointing nearly
vertically downward, forming a short tunnel that slopes
inward to a nest chamber. Each pair builds its own
nest, and the nest chamber is never joined to a
neighbour's. A new nest is never built away from the
main structure. Unusually for birds, these are not
nests utilized just for breeding but are also
dormitories that are used throughout the year.
These weavers are found only in southwest Africa.

Enlarging the colony

A new nest may have only 10 breeding pairs and
measure about 1m (3ft) across. As new pairs join
the colony, the roof is enlarged and new nests are
added from the bottom until as many as a hundred
families are living cooperatively in an enormous
structure that is 7.6m (25ft) long, 4.6m (15ft)
wide and 1.5m (5ft) high. The host tree sometimes
collapses under the weight of the nest, leaving the
colony stranded until the birds rebuild somewhere
else. However, nests often last for many seasons;
one nest is known to have been in use for over a
hundred years.

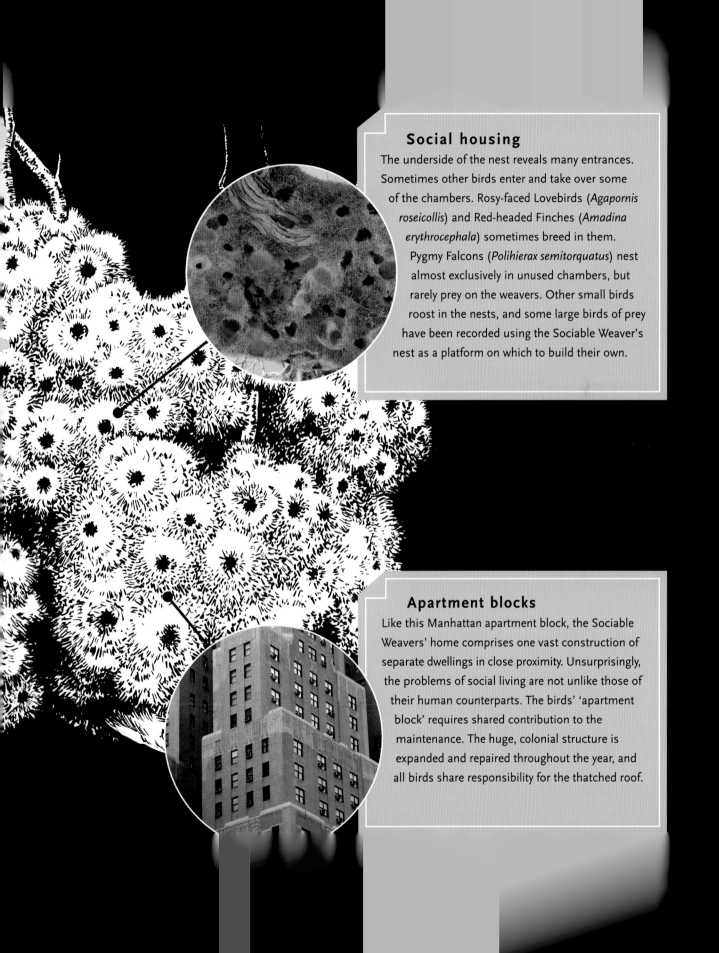

Social housing

The underside of the nest reveals many entrances. Sometimes other birds enter and take over some of the chambers. Rosy-faced Lovebirds (*Agapornis roseicollis*) and Red-headed Finches (*Amadina erythrocephala*) sometimes breed in them. Pygmy Falcons (*Polihierax semitorquatus*) nest almost exclusively in unused chambers, but rarely prey on the weavers. Other small birds roost in the nests, and some large birds of prey have been recorded using the Sociable Weaver's nest as a platform on which to build their own.

Apartment blocks

Like this Manhattan apartment block, the Sociable Weavers' home comprises one vast construction of separate dwellings in close proximity. Unsurprisingly, the problems of social living are not unlike those of their human counterparts. The birds' 'apartment block' requires shared contribution to the maintenance. The huge, colonial structure is expanded and repaired throughout the year, and all birds share responsibility for the thatched roof.

CASE STUDY
Rook

'Rookeries' are conspicuous features in the landscape across mid-latitudes in Europe and Asia. Eastern and far northern populations migrate, mostly to winter elsewhere in the breeding range. Rooks need tall trees for nests, facing open grassland and ploughed ground for feeding. They nest colonially and feed together on common ground, but each pair defends its nest and the area around it vigorously.

Colonies and roosts

Colonies are most often of several dozen pairs, but huge colonies of thousands of nests have been recorded in Hungary, the Netherlands and Scotland. All the adults and juveniles join birds of other colonies from July to February to form roosts of up to tens of thousands.

Nest building

At the start of the breeding season, the male alone builds a large part of the individual nest. First, a fresh or stolen twig is rested on a branch. When a few are resting successfully – many do fall – twigs held in the bill are then secured into place between existing twigs with upward and lateral pushes of the head. After 2–3 days, a platform is formed. The builder now stands on the platform and attaches twigs 3–60cm (1⅓– 24in) long in a dense ring around itself. Next, leaves, grass and earth are thrust into place between the twigs. The female forms a cup by treading down the bottom and forming the sides with her breast as she twists and turns. More soft materials, such as moss, feathers, leaves and even paper, complete the lining. The whole structure takes 1–4 weeks to complete.

Eggs and young

The female alone incubates an average of four eggs for 16–18 days. The young fledge in 30–36 days and are fed by the parents for another six weeks.

Classification

ORDER	Passeriformes
FAMILY	Corvidae
SPECIES	*Corvus frugilegus*
RELATED SPECIES	Crows, jays, jackdaws, ravens
NEST TYPE	Lined cup of sticks
SPECIES WITH SIMILAR NESTS	Crows and ravens
NEST SPECIALIZATION	Colonial rookery

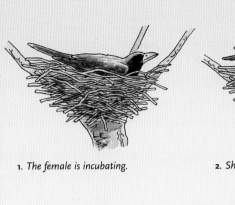

1. The female is incubating.

2. She senses a threat.

3. A male aggressor appears.

4. The female makes a threatening display.

5. Her mate chases the aggressor.

LEFT
STICK STEALING
A colony is filled with a constant 'cawing' sound. Early in the season, much of this is due to antagonistic behaviour as the pair try to drive away other males attempting to steal nest material – a regular feature of colony life. A female is incubating (1.). She will begin a threatening display (2.) if another male approaches (3.). The display consists of calling and fluffing her plumage aggressively (4.). The thief, however, often does dive in and steal a stick, before being chased away by the male mate (5.). (After Coombs, 1978.)

LEFT
TREETOP COLONY
This rookery in England clearly shows the scatter of nests in the treetops; most nests are single, but sometimes up to four will be clumped together. The average height for nests up a tree is 19m (62¹/₃ft). There are usually several nests in every tree, 0.5–2m (1²/₃–6²/₃ft) apart. Rookeries are busy from late February as the life-long pairs return to reclaim and repair last year's nests, and new pairs try to establish a site. The nest territory is defended until July.

CASE STUDY
Monk Parakeet

A gregarious species, the Monk Parakeet is the only parrot in the world that builds its own nest instead of using a cavity or hole in a tree. The Monk Parakeet is a common native of Argentina and neighbouring countries. This species is resident in dry, wooded country, mostly in scrub, palm groves and isolated clumps. It inhabits urban areas with planted trees, especially eucalyptus.

Nest building

Most commonly, 10–20 pairs of parakeets will build a large nest in a tree. It looks like a jumble of thorny sticks and twigs, but the manipulative skill of each pair enables them to build their own entrance at the side or from below. They then construct their individual nest chamber and line it with chewed twig debris. The colonial nest becomes an untidy, irregularly shaped structure measuring several feet across in each direction. Some colonies grow to be as big as 100 pairs and the nest is then the size of a small car. The naturalist W. H. Hudson found one in the early 1900s that weighed 200kg (440lb).

Neighbours

Empty 'apartments' in the nest are attractive nest sites for other birds, particularly the Yellow-billed Teal (*Anas flavirostris*), Brazilian Teal (*Amazonetta brasiliensis*) and Spot-winged Falconet (*Spiziapteryx circumcincta*). It is reported that none of the species pays strict attention to the others, not even the falconet. Although only about the size of the rightful owners, the falconet is known to prey on birds the size of adult parakeets as well as insects and lizards. When the second species moves in, the resident can do little except remain wary. The huge Jabiru (*Jabiru mycteria*) occasionally nests on top of the pile, although recent research suggests that the parakeets may in fact start building their nests at the bottom of the Jabiru's nest because of the support it provides. Whichever comes first, it does suggest an interesting association between the two species that may be beneficial to both.

Classification

ORDER	Psittaciformes
FAMILY	Psittacidae
SPECIES	*Myiopsitta monachus*
RELATED SPECIES	Other parrots, especially macaws and conures
NEST TYPE	Stick nest
SPECIES WITH SIMILAR NESTS	Some weaverbirds
NEST SPECIALIZATION	Large colonial stick nest

CASE STUDY
Florida Scrub Jay

Florida Scrub Jays live in extended family groups, building cup nests in oak scrub on the ancient dune systems. One family's territory can range between 2 and 20 ha (5 and 50 acres). The Florida Scrub Jay was formerly part of a widespread species that is now found only in southern Florida. The state and federal authorities have a vigorous protection and recovery plan for the species, which is still declining.

Classification

ORDER	Passeriformes
FAMILY	Corvidae
SPECIES	*Aphelocoma coerulescens*
RELATED SPECIES	Other scrub jays, Mexican Jay, Unicoloured Jay
NEST TYPE	Cup
SPECIES WITH SIMILAR NESTS	Many passerines
NEST SPECIALIZATION	Family groups

The group

Groups of jays range in size from a mated pair to large extended families of eight adults and several juveniles. Fledglings stay with the breeding pair as 'helpers'. This name was coined by Alexander Skutch, the ornithologist and specialist in tropical birds of the Americas, and is now widely used to describe the helping role of immature birds in a few species. The breeding male is dominant, then male helpers, followed by the breeding female, and lastly the female helpers. Helpers defend the territory throughout the year; share the lone duty of being sentinel on a good perch; mob predators; and feed nestlings and fledglings. Expansion of the birds' range is limited by the degradation and fragmentation of their specialized habitat. Juveniles that cannot find a suitable territory stay on with the family as helpers for up to five years.

Nest building

The nest is constructed low down in 1–3-m (3–10-ft) tall scrub oaks. An outer bulky basket of coarse oak twigs and other vegetation is formed with an outside diameter of 18–20cm (7–8in). Within this a 7.5–9-cm (3–3½-in) cup is shaped and lined with tightly wound palmetto or cabbage palm fibres. The rough structure shows well in the photograph.

Eggs and young

On average, a pair of jays will raise two fledglings per year. The presence of helpers improves the fledglings' chances of survival.

CHAPTER ELEVEN
Courts & Bowers

We have seen in previous chapters that courtship display is an important aspect of nest building. Among the bowerbirds this link between avian architecture and courtship reaches its fullest expression, but here the male builds a structure that does not serve as the traditional secure nest for eggs and young. It is instead a piece of statement architecture designed solely to attract females. The male bowerbirds are the garden architects, building and decorating ostentatious bowers, courts and lawns to charm and seduce.

Birds exhibit a fascinating range of courtship behaviours, from their many distinctive songs and calls to displays of fine feathers. The male European Robin (*Erithacus rubecula*) gains the attention of the female with a morsel of food. A pair of Western Grebes (*Aechmophorus occidentalis*) perform ritualized dances on water.

Among the bowerbirds, who constitute a single family found in New Guinea and Australia, the male's construction of display architecture is the defining courtship behaviour. He uses a range of techniques to build a lavishly decorated bower from sticks, plants and a vast range of decorative materials. This, together with vigorous dance movements and complex songs, is designed to attract as many females as possible. As far as the architecture is concerned, display is all. Once she has mated, a female will fly off alone to build a more traditional piece of nest architecture, a cup nest, in which to raise her young.

Dr Gerald Borgia, a biologist at the University of Maryland, believes there are at least five reasons for the building of a bower: (1.) In some cases, it functions as a device to attract females; (2.) it has an important role in controlling a male's display so the female can observe it from a comfortable position; (3.) it provides a stage for the male's performance; (4.) it helps orient the female; (5.) once inside the bower, it orients her to maximize the appeal of the male's display. In any population, there are only a few eligible males, so they get to mate with many females.

Ostentatious architecture has the advantage over eye-catching plumage because it is separate from the bird, making him less conspicuous to predators. But it does not guarantee success. It takes years of practice for a male to become a successful builder of bowers, and younger architects might fail to attract a single visitor.

RIGHT
GREAT BOWERBIRD
This male Great Bowerbird (Chlamydera nuchalis) *in Kakadu National Park, Northern Territory, Australia, stands in his avenue bower, which he builds to impress prospective mates.*

BLUEPRINTS

Bower Structures

Unlike the avian architecture featured previously, the blueprint for the courts and bowers is not concerned with construction of a secure nest for eggs and young. These structures are about display, although they are thought also to be designed to make a visiting female feel comfortable and protected from over-amorous males. The courts and bowers of this bird family can be divided into three types: 'stage', 'avenue' and 'maypole'. It may take a male seven years' practice before he succeeds in building architecture that attracts a mate.

FIG. I
AVENUE BOWER
The Fawn-breasted Bowerbird constructs a typical avenue. On a platform of sticks, he erects an avenue of firmly secured sticks, the inside of which he 'paints' with saliva and chewed vegetable pieces. One end or the other of the avenue is decorated with numerous, often colourful, objects. The Fawn-breasted's bower is characterized by a display of green berries.

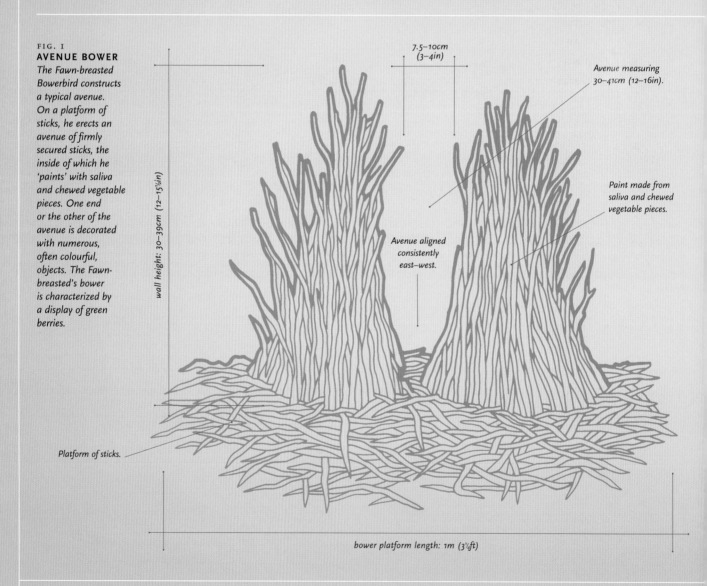

7.5–10cm
(3–4in)

Avenue measuring
30–41cm (12–16in).

Paint made from
saliva and chewed
vegetable pieces.

Avenue aligned
consistently
east–west.

wall height: 30–39cm (12–15¼in)

Platform of sticks.

bower platform length: 1m (3¼ft)

FIG. I FAWN-BREASTED BOWERBIRD BOWER

VARIETIES OF STRUCTURE

The simplest bower type is the stage built by two species, one of which is the Tooth-billed Bowerbird (*Scenopoeetes dentirostris*). More elaborate is the avenue bower built by eight species, including the Fawn-breasted Bowerbird (*Chlamydera cerviniventris*). The most complicated type is the maypole bower made by five species: the Golden Bowerbird (*Prionodura newtoniana*) produces a particularly elaborate example. Individual bowers vary hugely in size, so the measurements given here are from a particular example.

FIG. 2
MAYPOLE BOWER

Maypole bowers are constructed from a column of sticks built up around a sapling. The Golden Bowerbird goes one step further by building two towers bridged by a perch such as a vine stem. The bird displays on the perch above a few display objects, and between the towers that are generously decorated with lichen, flowers and berries. This bird is the smallest bowerbird, only 23cm (9in) long, but builds one of the largest bowers.

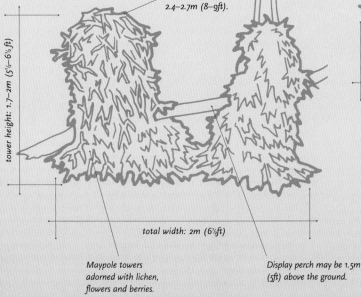

Towers may rise to 2.4–2.7m (8–9ft).

tower height: 1.7–2m (5½–6½ft)

total width: 2m (6½ft)

Maypole towers adorned with lichen, flowers and berries.

Display perch may be 1.5m (5ft) above the ground.

Fresh leaves cut with specially adapted bill.

Stage of leaves, shiny side up.

Careful placement of fresh leaves in the arena.

length: 1m (3¼ft)

FIG. 3
STAGE BOWER

The bower of the Tooth-billed Bowerbird hardly deserves that name compared with those of its relatives. The male simply clears the debris from a level area of the forest floor, and then arranges a number of large leaves with the pale or shiny side uppermost in the space to create the 'stage'. The stage bower is found only in the tropical forest of northeast Queensland, Australia.

FIG. 2 GOLDEN BOWERBIRD BOWER | FIG. 3 TOOTH-BILLED BOWERBIRD BOWER

MATERIALS AND FEATURES
Western Bowerbird Avenue

The Western Bowerbird (*Chlamydera guttata*) inhabits the outback of the Northern Territory and Western Australia. It builds an avenue bower that comprises two parallel walls of sticks about 60cm (2ft) long, 15cm (6in) apart and as high as 45cm (17¾). This architecture is constructed despite the bird being only 28cm (11in) long. The bulky walls of fine sticks, pushed into the ground and lined on the inside with dried grass, are 25cm (10in) thick at the base. The ground in front of each end of the bower and in the length of the avenue is usually decorated; in some bowers, the decoration extends in a wide ring all around.

Courtship display

As soon as a female arrives at the bower, the male stops arranging and tidying decorations and indulges in a frenzy of movement – running, jumping, short flights, picking up decorations, flinging them about, displaying lilac nape feathers that are usually hidden, and calling loud churrs and scolding notes. If the female is impressed, she will eventually crouch in the avenue and the two birds will mate.

Extravagant decoration

The male may display his building skills through the colour, arrangement or profusion of decorative materials. In sheep country, the Western Bowerbirds have a ready supply of grisly decoration – bones. A great quantity may be gathered, as shown in this bower in the Northern Territory. More than 1,000 bones have been found at one bower. Elsewhere, the male will collect coloured berries, pieces of glass, shiny pebbles, shells and shiny man-made materials that have included spoons, coins, foil and a glass eye.

Paintwork

The straw-coloured grass stems at the base of the avenue are painted brown by the bird. With a jabbing motion of the head, it wipes the grass with the sides of the bill or draws a grass stem between the mandibles. Microscopic investigation has shown the 'paint' to be a mixture of saliva and finely chewed particles of grass. In a rare example of birds using tools, some bowerbirds have used twigs to apply the paint for their interior decoration.

CASE STUDY

Tooth-Billed Bowerbird

The Tooth-billed Bowerbird or Toothbill is also known as the 'Stagemaker' because it constructs a courtship stage or arena. The bird itself is difficult to detect due to its dull brown upperparts and dirty white, brown-streaked underparts. It would be easily overlooked in the tropical rain forest of northeast Queensland, Australia, were it not for its very loud song and its bower.

The stage

By September or early October, at the beginning of the breeding season, the male chooses a level site on the forest floor and completely clears it of debris until he has a roughly circular arena, or stage; this will become his courtship site. Beside the stage, he needs a good perch from which to sing.

The set

The bird decorates the cleared ground with fresh leaves,which he cuts with his notched bill – hence the common name 'Toothbill'. He carefully arranges the leaves on the stage, with the lighter undersurfaces uppermost. Some Toothbills select leaves from only one tree; others cut a variety, often choosing leaves up to 25cm (10in) long and 15cm (6in) wide, although the bird is only 27cm (10½in) long. The tapering, 50-cm (19¾-in) long leaves of the wild ginger plant are particularly striking. Up to 20 large leaves may complete the set, but there may be as many as 100 small leaves. As the leaves wither, the male replaces them with new ones.

The main act

From the vantage point of his perch, the male pours forth a torrent of sound in a seemingly endless variety of calls. The final display and pairing have rarely been seen.

The finale

The nest is a frail, saucer shape made from thin, dry sticks built into thick foliage at the top of a sapling or in a mass of vine growing on a tree. Here, the female alone rears one or two young.

Classification

ORDER	Passeriformes
FAMILY	Ptilonorhynchidae
SPECIES	*Scenopoeetes dentirostris*
RELATED SPECIES	16 others in the family, and lyrebirds
NEST TYPE	Shallow, untidy cup
SPECIES WITH SIMILAR NESTS	Many passerines
NEST SPECIALIZATION	Stage bower

LEFT
TOOTHBILL ON SINGING STICK
Males sing throughout the day a great mixture of loud, harsh-sounding, clear, sweet and melodious calls. This Toothbill appears on his 'singing stick', overlooking his stage. He makes a loud, persistent series of calls. Several males can be found within hearing distance of each other. Clearly, they hold a 'singing contest', singing against each other in an effort to attract the most females.

LEFT
THE STAGE
The Toothbill's roughly circular stage, or arena, is about 1m (3ft) across. The males have rarely been observed displaying on this stage as other bowerbirds do in their bowers. The brighter, upturned leaves on cleared ground are a very noticeable feature of the forest floor. Together with the bird's torrent of song, they advertise the male to a female. The horizontal branch here makes an ideal 'singing stick' for this male. At other sites, the perch is sometimes 2–3m (6½–10ft) above the stage.

CASE STUDY
Satin Bowerbird

The male Satin Bowerbird builds an avenue bower, and demonstrates how particular colours are used in bower decoration to attract females. In contrast to the somewhat drab Toothbill, the male Satin Bowerbird is a deep, velvety blue-black. The female is a less spectacular grey-green above with brown crescent-shaped markings on pale grey underparts. Satin Bowerbirds are found only in the forests of Queensland and Victoria, Australia.

Avenue bower

The male uses sticks to make a rounded platform 2.5–4cm (1–1²/₃in) thick and 60–70cm (2–2¹/₄ft) in diameter. On this he attaches two parallel walls of sticks 30cm (1ft) long and 36cm (14in) high, spaced 10–12.5cm (4–4³/₄in) apart. Each wall is widened until the base is 12.5cm (4³/₄in) thick. The erect sticks tend to curve inwards at the top, forming an incomplete arch. Fine twigs are carefully attached to line the inside of the avenue, but the outside remains untidy. The avenue is aligned north–south, with a display platform at the sunny, north end. It is begun in July and deserted by December.

Decorating the bower

The male excels as a decorator. He is very colour conscious and collects mostly blue objects to decorate the northern display area. Some yellowish leaves are selected, but never anything red. The blue 15-cm (6-in) long tail feathers of the Crimson Rosella (*Platycercus elegans*), a parakeet, are much favoured. Satin Bowerbirds also 'paint' the inside of the bower to a height of about 15cm (6in). The paint is saliva mixed with crushed bark, ground-up charcoal and blue *Dianella* berries – the blue eventually turning black. The stems are often repainted until the layer is about 1–2mm (¹/₂₅in) thick. The male paints with a fibrous wad of material in his beak, using it like a brush or sponge; when it becomes dry, he makes a new wad.

Several bowers may be close together, allowing females to walk from one to another to choose a particularly well-constructed bower with many blue decorations, and select its builder as a mate.

Classification

ORDER	Passeriformes
FAMILY	Ptilonorhynchidae
SPECIES	*Ptilonorhynchus violaceus*
RELATED SPECIES	Other bowerbirds, lyrebirds
NEST TYPE	Shallow bowl
SPECIES WITH SIMILAR NESTS	Many passerines
NEST SPECIALIZATION	Avenue bower

SATIN BOWERBIRD AVENUE
This male has decorated the arena in front of the avenue. The mixture and brightness of the blue objects, especially the feathers, is clearly shown. The male displays by wing-waving, spreading his tail, picking up and waving a yellow object, such as a leaf, leaping to and fro and calling repeatedly. If the display is successful, the female remains in the bower and mating will take place.

1. *The bower displays snail shells and blue feathers.*

2. *Pieces of blue glass and plastic are used.*

3. *The male displays to the female in his bower.*

BOWER DECORATIONS
Snail shells are popular decorations in some bowers (1.). Bright blue is a favourite colour. Observers have seen blue paper, drink-bottle tops (2.), blue glass, blue plastic and even a blue toothbrush – all collected to help further the male's promiscuous lifestyle. A female, attracted to a bower by the male's song, will stand in the bower facing the displaying male before mating (3.). Males have been known to mate with two dozen females. The female builds a shallow, bowl-shaped nest that may be some distance from the bower.

CASE STUDY
Vogelkop Bowerbird

The Vogelkop Bowerbird is a maypole builder that lives in the mountains of the Vogelkop Peninsula, Western New Guinea, Indonesia. Among the maypole builders there is an inverse correlation between the colourfulness of their plumage and the complexity of their bowers: the duller the plumage, the more extravagant the bower. The Vogelkop Bowerbird is the least colourful and builds the most elaborate bower of all.

Maypole bower

The bird makes the maypole structure by weaving a tower of sticks and orchid stems around one or two saplings stripped of leaves. The maypole is then covered over with a canopy of interlocking sticks that may be supported by one or more outlying saplings. The canopy would be circular if there wasn't a cavernous entrance on one side. The maypole bower may be about 2m (6½ft) tall and have an even greater diameter, which is more astonishing when one considers that the male is only 25cm (10in) long.

Decorating the bower

The Vogelkop covers the floor of the bower and the 'garden' outside with moss, and decorates this with many objects. Unlike the Satin Bowerbird, he appears to collect these for their novelty value rather than for their colour. The birds have plenty of choice, and Vogelkop Bowerbirds have individual decorative preferences. Coloured fruits, black beetle wing-cases, black bracket fungi, deer droppings and flowers are all common decorations. The bird gathers the collected items into piles of the same kind and colour, and places them in different parts of the bower. Males are very house-proud and constantly attend their bowers, replacing old items with spectacular new ones.

Bower theft

Neighbours, who may be less than 200m (660ft) away, are clearly in competition for mates because raids are frequent, even though close neighbours' decorations are often completely different.

Classification

ORDER	Passeriformes
FAMILY	Ptilonorhynchidae
SPECIES	*Amblyornis inornata*
RELATED SPECIES	Other bowerbirds, lyrebirds
NEST TYPE	Simple cup
SPECIES WITH SIMILAR NESTS	Many passerines
NEST SPECIALIZATION	Maypole bower

LEFT
VOGELKOP MAYPOLE
The bird's dull plumage is in great contrast to the brightness of his display of red berries and shiny black beetle wing-cases. Females visit the bowers to assess the quality of the male's building and to hear his astonishing song repertoire, both of which are indicative of his potential as a mate. The full song is performed hidden behind the maypole. The female alone builds the nest and rears the young.

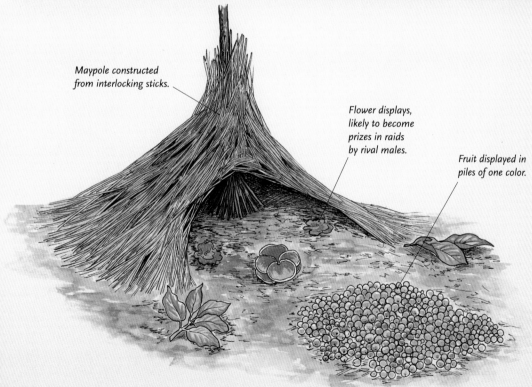

Maypole constructed from interlocking sticks.

Flower displays, likely to become prizes in raids by rival males.

Fruit displayed in piles of one color.

LEFT
BOWER GARDENING
The work is not over when the bower appears to be finished. The male spends up to nine months gardening in front of his bower. Every day, the gardener throws out faded or rotten decorations; if white fungus starts to grow on the deer droppings, he trims it off. He then goes out to find fresh decorations. Ornithologists have established that a male spends 4–7 years practicing bower-making and observing the work of older birds before he eventually breeds.

CHAPTER TWELVE

Edible Nests & Food Stores

A number of bird species use specialist skills to create and exploit food stores that can provide them with a steady supply of nutrients. They construct and use pantries, granaries and orchards of various designs. There are also two species that produce a natural nest-building material which has become a significant human food resource.

The Edible-nest Swiftlet (*Aerodramus fuciphagus*) and Black-nest Swiftlet (*A. maximus*), small swiftlets of southeast Asia, are the only birds in the world that manufacture their own nest material. Their nests have been collected and sold for human consumption for centuries. It is recorded that in Roman times the physician Andromachus made a medicine for Emperor Nero from these nests. They have long been a luxury food for people, often dissolved in chicken or mutton broth to add taste. Today, they are in particular demand in China to make bird's-nest soup.

Even as late as the mid-nineteenth century, scientists were not certain how these birds created the material for their nests. Different theories proposed that the swiftlets collected the foam formed by the sea lashing against rocks; the sperm of whales; or fish spawn floating on the sea. It is now known that the birds construct their edible nests from threads of their own solidified saliva.

Other species are noteworthy for their construction of food stores. Birds work hard to nourish themselves. Whether they eat insects, other birds or mammals, fish or seeds, it takes up a great deal of their time and energy.

Thus, the storing of hard-won food aids survival. The Eurasian Jay (*Garrulus glandarius*) collects acorns in autumn and buries them; many are found again later in the winter when food is becoming naturally scarce. Several members of the tit family (Paridae) in Europe and North America store food – seeds, nuts and even immobilized caterpillars – especially under the bark of trees; but also, in milder climes, in grass tussocks and under moss.

Shrikes (*Lanius* spp.) are famous for their 'pantries' of large insects, small mammals or lizards that are impaled on the thorns of trees. The Acorn Woodpecker (*Melanerpes formicivorus*) builds a spectacular 'granary' of individual holes, as shown above. The bird uses these holes to lodge and store acorns. Tits and shrikes store short term, for days or hours. The Acorn Woodpecker, however, is exceptional for the storage of a large amount for survival over several months.

Other species tend 'orchards' where they drill patterns of 'harvest wells' in the trees. The sapsucker woodpecker of North America defends a territory in which there are trees that provide a rich source of sap. The bird has mastered the task of making sap flow abundantly from these trees, which it laps up with its brush-like tongue.

RIGHT
WILLIAMSON'S SAPSUCKER
A male Williamson's Sapsucker (Sphyrapicus thyroideus) *at a nest hole in an aspen tree in Rocky Mountain National Park, Colorado.*

BLUEPRINTS
Edible Nest & Food Store Structures

The edible nest blueprint takes the form of threads of solidified saliva that are built up gradually in a horseshoe shape. These nests are produced by Edible-nest and Black-nest Swiftlets (*Aerodramus fuciphagus* and *A. maximus* – small, white-rumped swifts found only in Malaysia and Indonesia, breeding mostly on offshore islets. Food storage structures are created by woodpeckers known as sapsuckers and the Acorn Woodpecker (*Melanerpes formicivorus*).

FIG. 1
BLACK EDIBLE NEST
The nests of the Black-nest Swiftlet are found in huge colonies in caves, many on islands in Southeast Asia. As a source of food, their nests are much less valuable because the saliva is mixed with feathers and vegetable matter. This has to be separated from the saliva strands, which are the soup's raw material, and this is a relatively expensive process.

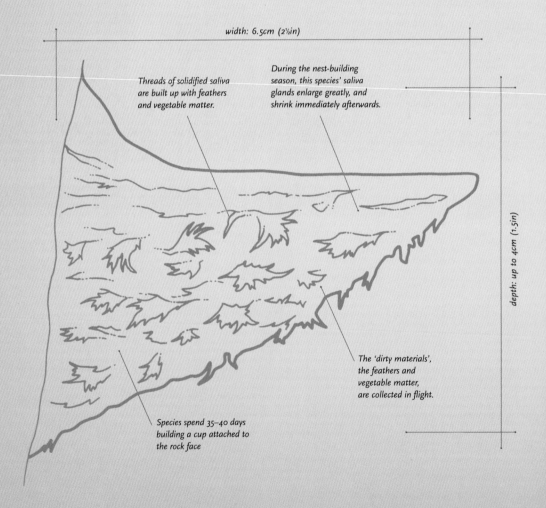

width: 6.5cm (2½in)

Threads of solidified saliva are built up with feathers and vegetable matter.

During the nest-building season, this species' saliva glands enlarge greatly, and shrink immediately afterwards.

depth: up to 4cm (1.5in)

The 'dirty materials', the feathers and vegetable matter, are collected in flight.

Species spend 35–40 days building a cup attached to the rock face

FIG. 1 BLACK-NEST SWIFTLET NEST

VARIETIES OF STRUCTURE

The Black-nest Swiftlet mixes saliva with other materials, whereas the Edible-nest Swiflet builds purely with saliva. Swiftlets are the only birds, apart from those species that build completely or partly with mud, to mould the nest material into shape. Sapsuckers construct their own food supply by tapping a variety of trees for the nutritious, summer phloem sap, which contains sugars. The Acorn Woodpecker creates a 'granary' by drilling holes in dead tree branches or even telephone poles to store acorns.

Holes are drilled through the bark and into the sapwood, providing food for three days.

When one line of holes dries up, a new line is created above.

holes: 1.25cm (½in) across

Among the most easily spotted 'wells' are those in birch trees.

Sap flows from the hole and attracts insects – both good food sources.

FIG. 2
WHITE EDIBLE NEST

The nest of the Edible-nest Swiftlet is constructed entirely from the bird's saliva. It is stuck to the rock face in a huge cave or inside a specially constructed building in Southeast Asia.

A zoologist recorded that in one night a single bird laid one strip of saliva to the nest that was 4.5cm (1¾in) long, 8mm (⅓in) high and 2.5mm (⅒in) thick. In some regions, the

bracket-shaped nest is harvested to make bird's-nest soup. The birds nest in huge colonies, and in the darkness of the caves they navigate by echolocation.

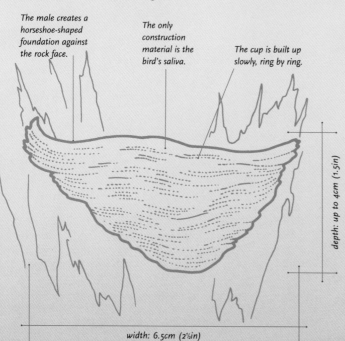

The male creates a horseshoe-shaped foundation against the rock face.

The only construction material is the bird's saliva.

The cup is built up slowly, ring by ring.

depth: up to 4cm (1.5in)

width: 6.5cm (2½in)

FIG. 3
A SAPSUCKER'S FOOD STORE

The sapsucker woodpeckers of North America are well known for the rings of holes that they drill in certain trees. The Yellow-bellied Sapsucker (Sphyrapicus varius)

drinks the sap that flows from each hole, and eats the insects that are attracted to it. Sapsuckers have an anticoagulant in their saliva that prevents the sap from congealing.

FIG. 2 EDIBLE-NEST SWIFTLET NEST | FIG. 3 YELLOW-BELLIED SAPSUCKER WELL

MATERIALS AND FEATURES
Edible-Nest Swiftlet Nest

Edible nests, which are used to produce the famous bird's-nest soup, are made by two species found in Southeast Asia, the Edible-nest Swiftlet (*Aerodramus fuciphagus*) and the Black-nest Swiftlet (*A. maximus*).

The former's nest is completely created from the bird's saliva. The salivary glands are greatly enlarged during the breeding season. First, the bird flies at a cave wall to deposit drops of saliva to form a horseshoe-shaped foundation. Then it builds up the base to form a shallow cupped nest. The bird builds up the nest in concentric layers of salivary 'cement'. The layers often show as horizontal marks on the nest cup's outer surface.

Each nest weighs about 14g (½oz), produced by a small, black swiftlet that weighs only 14–18g (½–⅔oz), and is only 11.5cm (4½in) long. The nest is soft at first but hardens with age and deteriorates after a few months.

Edible nest colonies
Several thousand birds may nest in one colony. The birds nest very close to the cave ceiling – for example, in the famous Niah Caves of Sarawak, Malaysia, the nest harvesters' scaffolding is 90–120m (300–400ft) high. They prefer to nest in dark sites. The birds nest in close groupings, although individual nests are usually constructed no closer than 5cm (2in) apart.

Lattice structure

The horseshoe-shaped foundation acts as a bracket, attaching the nest to the rock or wall. It is an intricate lattice of white, hardened strands of salivary laminae cement, which extends across the space from the raised wall of the front of the cup to the vertical surface. It is not attached to a projection. The saliva's adhesive nature glues it to the nest site. The male builds the nest over a period of about 35 days.

Urban nesting sites

Since the 1990s, a huge nest-farming industry has grown up in Indonesia, especially in northern Sumatra. It developed after urban dwellers noted that swiftlets, known as 'walet', sometimes nested in the upper floors of their homes. The growth of farming has led to the creation of 'walet hotels', with their cave-like entrances. The 'cave' entrance is just visible in the photograph, top right. In Kisaran, a Sumatran town of 70,000 people, there are 300 walet hotels in the city centre alone. The nests provide valuable extra income.

CASE STUDY
Acorn Woodpecker

The Acorn Woodpecker's speciality is the construction of a food store. Resident in the oak forests of the southwestern United States south to Colombia, these birds are often found in family groups. A new pair will establish a territory, and as the family grows, the group may include as many as seven males, three females and ten non-breeding helpers. Together, this family group defends the territory and its trees for food, storage 'granaries', nest holes and roost holes.

The nest hole

Acorn Woodpeckers nest in an excavated cavity in a large tree and may reuse the hole for several years. Usually, the nest contains the eggs of one female. If another female lays there she will often destroy any eggs she finds in the nest. The group share the incubation and the feeding of the young. Subordinate members of a group that reach breeding condition will find another group to prevent inbreeding.

Food store construction

Although the birds will feed on insects and their larvae, their main source of food, especially in winter, is acorns. To store the acorns, they construct a 'granary' by drilling acorn-size holes in a dead tree branch or a utility pole. In the fall, the group stores an acorn in each hole and defends the cache against possible thieves, such as Steller's Jays (*Cyanocitta stelleri*), California Scrub Jays (*Aphelocoma californica*) and Lewis's Woodpeckers (*Melanerpes lewis*). The 'granary' is used over several years and may eventually have as many as 50,000 holes, as generations of the birds continue to use a successfully defended tree. Each territory on average has two such trees, but as many as seven have been recorded. The holes are narrower at the mouth than further in so that an acorn can be driven in, pointed end first, and will not fall out.

Additional food storage

Acorn Woodpeckers also store pine seeds and nuts, such as almonds, walnuts and pecans, in fissures in trees and cracks in posts and walls.

Classification

ORDER	Piciformes
FAMILY	Picidae
SPECIES	*Melanerpes formicivorus*
RELATED SPECIES	Other woodpeckers, piculets, wrynecks
NEST TYPE	Excavated hole
SPECIES WITH SIMILAR NESTS	Other woodpeckers, kingfishers, bee-eaters
NEST SPECIALIZATION	Food stores

LEFT
GRANARY TREE
Acorn Woodpeckers can drill quickly. On one April day, a female was observed drilling 286 holes in a pine trunk in 9 hours 27 minutes. These three birds are at a 'granary tree' or 'nut pole', where there are many stored acorns, but no grain, despite the name. The two upper birds are giving a spread-winged greeting display. When a bird wants to eat an acorn, it carries the acorn to an 'anvil' — a crack, in which the acorn fits securely. The bird then splits the acorn by striking it with its bill. Any pieces not eaten are put back into store.

CASE STUDY
Yellow-Bellied Sapsucker

The Yellow-bellied Sapsucker feeds on sap from trees, and its creation of an 'orchard' of sap-harvesting 'wells' ensures a plentiful supply for the family. This sapsucker is in fact a woodpecker found across North America in a band from eastern Alaska to Newfoundland and New England. Unusually for a woodpecker, it is a migrant, wintering in the southeastern states, Mexico and through Central America.

The nest site

The male chooses and defends a territory, selects the nest site and excavates a typical woodpecker cavity 15–35.5cm (6–14in) deep, which takes 2–4 weeks. Both birds care for the usual brood of 5–6 young in a breeding season that lasts up to 11 weeks long.

The sapsucker 'orchard'

Sapsuckers defend a territory – an 'orchard' – in which there may be several trees that are good providers of sap. The birds feed on a large range of trees but in any area will have favourites, such as birches, sugar maples, Scots pines, aspens, juneberry or willows. In particular, sapsuckers seek the summer phloem sap, the sticky fluid produced in the leaves that flows down to other parts of the tree. They drill lines of 'harvest wells' then lap up the sap with their long tongues, which are lined each side with brush-like edges that help to gather in the fluid. As the tree seals the wound, the sap dries up, so the sapsucker drills another line of wells above the old ones; this process is repeated up the tree.

Well visitors

Because sapsucker wells maintain the flow of sap, other creatures take advantage of this food source, including many other birds, bats, squirrels and porcupines. It is thought that Ruby-throated Hummingbirds (*Archilochus colubris*) – see pages 60–61 – time their springtime arrival in the north to coincide with the peak sapsucker drilling time. The sap contains similar amounts of sugar and other nutrients to those in the flowers on which the hummingbirds normally feed.

Classification

ORDER	Piciformes
FAMILY	Picidae
SPECIES	*Sphyrapicus varius*
RELATED SPECIES	Other woodpeckers, wrynecks, piculets
NEST TYPE	Self-excavated tree cavity
SPECIES WITH SIMILAR NESTS	Woodpeckers, kingfishers, bee-eaters
NEST SPECIALIZATION	Food well

LEFT
SAPSUCKER'S HARVEST WELL

Here, a male is tending his well by Lake Nettie, Michigan. The birds drill holes in a horizontal ring, over 3–4m (10–13ft) high. The holes are about 6mm (¼in) wide and deep, going into the bark and through the cambium (inner bark), and then another 3mm (⅛in) into the sapwood. A new hole can provide rich food for about three days, so the sapsuckers 'farm' the produce throughout the season. They cluster the holes to induce the accumulation of sap in the bark, ready for harvesting later.

Resources

BOOKS

Armstrong, Edward A., *The Wren* (Collins, 1955)

Attenborough, David, *The Life of Birds* (BBC Books, 1998)

Bannerman, David, *The Birds of the British Isles*, vol. 5 & 11 (Oliver & Boyd, 1956)

Bent, Arthur Cleveland, *Life Histories of North American Birds*, 21 volumes (Dover Publications, 1919–1968)

Beruldsen, Gordon, *A Field Guide to Nests and Eggs of Australian Birds* (G. & E. Beruldsen, 1980)

Campbell, Bruce and Ferguson-Lees, James, *A Field Guide to Birds' Nests* (Constable & Co. Ltd, 1972)

Chaffer, Norman, *In Quest of Bower Birds* (Rigby, 1984)

Collias, Nicholas E. and Collias, Elsie C., "An Experimental Study of the Mechanisms of Nest Building in a Weaverbird" in *The Auk* vol. 79: pp. 568–595 (1962)

Collias, Nicholas E. and Collias, Elsie C., *Nest Building and Bird Behavior* (Princeton University Press, 1984)

Coombs, Franklin, *The Crows* (Batsford, 1978)

Cramp, S. *et al.* (Editors), *The Birds of the Western Palearctic*, volumes 1–9 (Oxford University Press, 1977–1994)

Davies, S. J. J. F., "The Nest-Building Behaviour of the Magpie Goose" in *Ibis* vol. 104: pp. 147–157 (1962)

Diamond, Jared, "Bower Building and Decoration by the Bowerbird *Amblyornis inornatus*" in *Ethology* vol. 74: pp. 117–204 (1987)

Dunning, Joan, *Secrets of the Nest: The Family Life of North American Birds* (Houghton Mifflin, 1994)

Durrell, Gerald, *The Drunken Forest* (Penguin, 1956)

Forbush, Edward Howe (revised by John Richard May), *A Natural History of American Birds* (Bramhall House, 1955)

Frith, H. J., *The Malleefowl* (Angus & Roberston Limited, 1962)

Goodfellow, Peter, *Birds as Builders* (Arco Publishing, 1977)

Goth, Ann and Booth, David T., "Temperature-Dependent Sex Ratio in a Bird" in *Biology Letters*, The Royal Society online (2005)

Hansell, Mike, *Bird Nests and Construction Behavior* (Cambridge University Press, 2000)

Harrison, Colin, *Nests, Eggs and Nestlings of North American Birds* (Collins, 1978)

Harrison, Colin and Castell, Peter, *Bird Nests, Eggs & Nestlings of Britain & Europe* (HarperCollins, revised 1998)

Haverschmidt, F., *The Life of the White Stork* (E. J. Brill, 1949)

Herrick, Francis H., "Nests and Nest Building in Birds" in *Journal of Animal Behaviour*, vol. 1: pp. 159–192, 244–277, 336–373 (1911)

Howard, J., "Sticky Solutions" in *New Scientist, Inside Science*, No. 102, 1–4 (1997)

Hume, Robert, *The Common Tern* (Hamlyn, 1993)

Jameson, William, *The Wandering Albatross* (Morrow, 1958)

Jaramillo, Alvaro and Burke, Peter, *New World Blackbirds – The Icterids* (Christopher Helm, 1999)

Long, Ashley M., Jensen, W. E., and With, K. A., "Orientation of Grasshopper Sparrow and Eastern Meadowlark Nests in Relation to Wind Direction" in *The Condor* vol. 111: pp. 395–399 (2009)

Maclean, G. L., "The Breeding Biology of the Double-Banded Courser" in *Ibis* vol. 109: pp. 556–569 (1967)

Moreno, J., Soler, M., Møller, A. P., and Linden, M. "The Function of Stone Carrying in the Black Wheatear, *Oenanthe leucura*" in *Animal Behaviour* 47, pp. 1297–1309 (1994)

Ogilvie, Malcolm, *Grebes of the World* (Bruce Coleman Books, 2002)

Phillips, W. W. A., *Birds of Ceylon,* vol. 2 (Associated Newspapers of Ceylon, 1952)

Rankin, Niall, *Haunts of British Divers* (Collins, 1947)

Reid, J. M., Cresswell, W., Holt, S., Mellanby, R. J., Whitfield, D. P., and Ruxton, G. D., "Nest Scrape Design and Clutch Heat Loss in Pectoral Sandpipers" in *Functional Ecology* vol. 16, issue 3 (2002)

Rennie, J., *The Architecture of Birds* (G. Cox, 1831)

Richardson, Frank, "Breeding and Feeding Habits of the Black Wheatear of Southern Spain" in *Ibis* vol. 107: pp. 1–16 (1965)

Roseberry, J. L., and Klimstra, W. D., "The nesting ecology and reproductive performance of the Eastern Meadowlark" in *Wilson Bulletin* vol. 82: pp. 243–267 (1970)

Rowley, I., "The Use of Mud in Nest-Building" in *Ostrich,* Suppl. 8: pp. 139–148 (1970)

Saunders, Howard, *Manual of British Birds* (Gurney & Jackson, 1899)

Scheithauer, Walter, *Hummingbirds* (Thomas Y. Crowell & Co., 1967)

Simmonds, Calvin, *Private Lives of Garden Birds* (Storey Publications, 2002)

Simmons, K. E. L., "Studies on Great Crested Grebes" in *The Avicultural Magazine*, vol. 61: pp. 235–253 (1955)

Simon, Jose Eduardo and Pacheco, Sergio, "On the Standardization of Nest Descriptions of Neotropical Birds" in *Revista Brasileira de Ornitologia* vol. 13 (2): pp. 143–154 (2005)

Skutch, Alexander F., *The Life of the Hummingbird* (Vineyard Books, 1974)

Skutch, Alexander F., *Parent Birds and Their Young* (University of Texas Press, 1976)

Skutch, Alexander F., *Life of the Woodpecker* (Ibis Publishing Company, 1985)

Skutch, Alexander F. *Helpers at Birds' Nest – a Worldwide Survey of Cooperative Breeding and Related Behavior* (University of Iowa Press, 1987)

Sparks, John and Soper, Tony, *Penguins* (Taplinger Publishing Company, 1967)

Sterry, Paul and Small, Brian E., *Birds of Eastern North America* (Princeton University Press, 2009)

Sterry, Paul and Small, Brian E., *Birds of Western North America* (Princeton University Press, 2009)

Stokes, Donald and Stokes, Lillian, *Hummingbird Book – A Complete Guide to Attracting, Identifying and Enjoying Hummingbirds* (Little, Brown and Company, 1989)

Thomson, A. Landsborough (Editor), *A New Dictionary of Birds* (McGraw-Hill, 1964)

Tucker, Priscilla, *The Return of the Bald Eagle* (Stackpole Books, 1994)

Turner, Angela and Rose, Chris, *Swallows and Martins* (Houghton Mifflin Company, 1989)

Wauer, Roland H., *The American Robin* (University of Texas Press, 1999)

Welty, Joel Carl, *The Life Of Birds* (W. B. Saunders Company, 1964)

Wood, J. G., *Wonderful Nests* (Longmans, Green & Co., 1892)

WEBSITES

In this book the birds' names are as found in the IOC World Bird List version 3.1 at http://www.worldbirdnames.org/ and the International English name in the British list of the British Ornithologists' Union at: http://www.bou.org.uk/

Arkive
Multimedia guide to the world's endangered species.
http://www.arkive.org/birds/

Avian Reproduction: Nests
Useful website with information on birds' nests and construction.
http://people.eku.edu/ritchisong/birdnests.html

Avibase – The World Bird Database
An extensive database information system about all birds of the world.
http://avibase.bsc-eoc.org/avibase.jsp?pg=home&lang=EN

BirdLife International
The world's largest partnership of conservation organizations.
http://www.birdlife.org/index.html

The British Trust for Ornithology
The UK's leading ornithological research organization with an international reputation for survey work.
http://www.bto.org/

Handbook of Birds of the World
A monumental, much-praised series planned in 16 volumes.
http://en.wikipedia.org/wiki/Handbook_of_the_Birds_of_the_World

The Royal Society for the Protection of Birds
The UK's premier bird conservation and protection society with over a million members, and many bird reserves.
http://www.rspb.org.uk/

World Bird Gallery
An image gallery of over 5,000 species.
http://www.camacdonald.com/birding/Sampler.htm

Glossary

altricial of young birds, helpless and requiring parental assistance when hatched.

arboreal living in or among trees.

bower a construction on the ground built by males of one Australasian family of birds; the male sings and displays here to attract a mate.

brood the number of young birds produced at one hatching.

clutch the number of eggs laid by a particular bird, e.g., a Robin, or laid in a single nest, e.g., Ostrich.

cock nest a nest built by the male bird that may be chosen by the hen as the one in which to raise a brood.

convergence evolutionary similarity between unrelated or distantly related forms as a result of adaptation to a similar mode of life.

coverts the feathers that cover above and below the bases of the tail and primary feathers (see below), and form the contour of the bird in those areas.

cuticle a protective outer layer.

eyrie the nest of an eagle or other large bird of prey.

feral of wildlife, existing in a wild state after having been domesticated.

fledge to grow feathers and become able to fly.

fledgling a young bird that has become able to fly from the nest.

gregarious of birds, living together in flocks.

icterid the common family name (such as sparrow or thrush) for members of the family Icteridae that contains American blackbirds, orioles, grackles, cowbirds, caciques and oropendolas.

incubate to provide heat to eggs for the development of the embryo.

nidicolous remaining in the nest until fledging.

nidifugous leaving the nest immediately or soon after hatching.

passerine of birds belonging to the Order Passeriformes, which are characterized by the perching habit, e.g., warblers, finches, thrushes, larks and crows. Popularly called 'songbirds'.

order in the classification of wildlife, the primary category of a class. It is a grouping of related families, e.g., Anseriformes – swans, geese and ducks.

pied having markings of black and white, as in the Magpie.

primary feathers the wing feathers often called the 'flight feathers', borne on the 'hand bones', as opposed to the 'secondaries' on the arm, and thus forming the bird's distinctive wing tip.

precocial describing young birds that hatch with eyes open, are covered with down, are capable of walking just after hatching, and leave the nest within the first day or two.

raptor word for 'bird of prey', a group that includes eagles, harriers, hawks and falcons.

ratite a flightless bird such as an Ostrich or Emu; a ratite's sternum lacks a keel for wing muscles.

relict species species in isolated or discontinuous populations that appear to represent a former much wider distribution.

retort description of a nest that resembles the shape of a glass vessel with a long neck on a round bulb base.

rookery a colony of nesting birds, from the name of the colony of the European Rook, but also applied to a colony of penguins.

roost the place where birds rest, for example, a particular branch for a pigeon, or a wood for starlings, or a sandbank away from predators for waders.

scrape the hollow made in soft ground to hold the clutch of eggs laid by ground-nesting birds, such as plovers or terns.

scrub vegetation consisting of bushes, stunted trees and other vegetation, often in arid areas.

spoil waste material dug out when excavating.

stage the area where a male bird displays to attract a female.

wader a shorebird, such as a plover, sandpiper, snipe or godwit.

Index

Page numbers in boldface type refer
to pages containing illustrations.

AUTHOR ACKNOWLEDGEMENTS

After a lifetime of birdwatching, I'm first of all grateful to all the friends who have watched with me, shared experiences from around the world, and have helped to keep the enjoyment of it all fresh; and also to all the host of bird-watchers and scientists whose writings describe the nests I would love to see but never will.

I'm very grateful to Ivy Press for the invitation to write *Avian Architecture*, thus allowing me to relive the schoolboy excitement of 'bird nesting' and now, as a grown-up birdwatcher, to reread my *Birds as Builders* and write about nests again.

At Ivy Press, I'm particularly grateful to Steve Luck, who did the initial editing; to Coral Mula for her illustrations, and her patience in making alterations and improvements; to Prof. Mike Hansell, who worked hard to keep my science on track and up to date; and to Caroline Earle, whose final control of editing made sure I kept writing about what birds build and stopped me from wandering off, getting interested in other topics that were not close enough to the main subject.

My sincere thanks goes to them all for helping me write about such a fascinating subject, and it leaves me hoping that you, the reader, will appreciate the wonders of bird architecture as much as I do.

PICTURE CREDITS

The publisher would like to thank the following individuals and organizations for their kind permission to reproduce the images in this book. Every effort has been made to acknowledge the pictures; however, we apologize if there are any unintentional omissions:

Alamy/Arco Images GmbH: 6; The Natural History Museum: 16, 44, 80, 92, 104; Peter Arnold: 26b; Pat Bennett: 29t; Geraldine Buckley: 40bl; Vasiliy Vishnevskiy: 39br; Frank Blackburn: 51t; William Leaman: 53b; Juniors Bildarchiv: 55; Tom Uhlman: 61t; Birdpix: 69; Tom Uhlman: 74l; The Art Archive: 128; Gomersall: 129b; Blickwinkel: 130; SilksAtSunrise: 153.
Steven Aronoff of Bellingham, Washington, USA: 139b.
Brian Barnes: 23.
Stephen Michael Barnett: 89t.
Leo Berzins: 2, 132.

Bridgeman Art Library/The Natural History Museum: 138, 140, 142.
Kensiong Chew: 149b.
Corbis/D. Robert & Lorri Franz: 11; Vasily Fedosenko/Reuters: 43b; Ron Austing/Frank Lane Picture Agency:60l; Gary W. Carter: 68; Tom Bean: 93; Joe Mcdonald: 100bl; Martin Harvey: 115tl.
Will Craven: 38.
Hanne & Jens Eriksen/www.BirdsOman.com: 14bl.
FLPA/Adri Hoogendijk: 77tl; Michael & Patricia Fogden: 98b; David Hosking: 105b; T S Zylva: 106, 148l; Gerry Ellis: 109; Roger Tidman: 119b; ImageBroker: 121t; Pete Oxford: 123; Michael & Patricia Fogden: 136b; Michael Gore: 139t; Thomas Marent/ Minden Pictures: 141t; Konrad Wothe/ Minden Pictures: 149t.
Fotolia: 31b, 63br, 75t; Johan Swanepoel: 9; Mycteria: 10tr; Diane Stamatelatos: 19b; D. Robert & Lorri Franz: 22tr; Florian Andronache: 46; John P. Marechal: 56; Aberenyi: 61b; RTimages: 62bl; Rusty Dodson: 75b; Feng Yu: 84; Duncan Noakes: 94; Charles Taylor: 99t; Pippa West: 99b; David Biagi: 108; Sebastien Closs: 113b; Elenathewise: 122; John Mccabe: 127b; Flucas: 137b.
Getty Images/De Agostini: 28l, 118, 120, 150; Mark Van Aardt: 127t; Belinda Wright: 133.
Johann & Lizet Grobbelaar/www.grobimages.co.za: 27t; 27b.
iStockphoto/Terry Wilson: 7; Steven Heap: 17t; Missing35MM: 32; AFhunta: 34tr; Peeter Viisimaa: 50b; Borre Heitmann Holmeslet: 70; James Metcalf: 74l; Ines Gesell: 88; Andre Maritz: 126b; Stephen Nash: 144.
Enrique F. Larreta: 113t.
Frederico Crema Leis: 83.
Michael Lyncheski www.mllpix.com: 103t.
Haim & Amanda Mayan: 15br.
Jenny Miners: 89b.
Nature Picture Library/Mike Potts: 15tr; Steve Knell: 20; Simon King: 21; Rolf Nussbaumer: 35; Tom Vezo: 47; Michel Poinsignon: 51b; Tony Heald: 54; Mike Read: 71; Hermann Brehm: 91tl; Jose B. Ruiz: 112b; Chris Packham: 131; Dave Watts: 136b; Richard Kirby: 143t; Marie Read: 151.
Photolibrary/Joachim Hiltmann: 45; Gunter Ziesler: 43t; Winifried Schafer: 67; Ralph Reinhold: 79b; Michel Poinsignon: 81; Nigel Dennis: 82; Horst Jegen: 85; GTW: 95; Roger Brown: 117b; Rolf Nussbaumer: 145.
Photoshot/Woodfall Wild Images: 33; Tracy Underwood: 57; NHPA: 107.
Jerry Schoen: 65b.
Lui Weber: 137t.